外教社 英汉·汉英百科词汇手册系列

总主编 庄智象

英汉·汉英
物理学词汇手册

An English-Chinese Chinese-English Glossary of Physics

编者 张军鹰

外教社 上海外语教育出版社
SHANGHAI FOREIGN LANGUAGE EDUCATION PRESS

图书在版编目(CIP)数据

英汉·汉英物理学词汇手册/张军鹰编.—上海:
上海外语教育出版社,2009
(外教社英汉·汉英百科词汇手册系列)
ISBN 978-7-5446-1572-3

Ⅰ.英… Ⅱ.张… Ⅲ.物理学—词汇—手册—英、汉
Ⅳ.04-62

中国版本图书馆CIP数据核字(2009)第193077号

出版发行:上海外语教育出版社
(上海外国语大学内) 邮编:200083
电 话:021-65425300(总机)
电子邮箱:bookinfo@sflep.com.cn
网 址:http://www.sflep.com.cn http://www.sflep.com
责任编辑:贺 敏

印 刷:上海叶大印务发展有限公司
经 销:新华书店上海发行所
开 本:787×1092 1/32 印张8.875 字数385千字
版 次:2010年1月第1版 2010年1月第1次印刷
印 数:3 100册

书 号:ISBN 978-7-5446-1572-3 / O·0008
定 价:19.00元

目 录

外教社英汉·汉英百科词汇手册系列

总主编：庄智象

英汉·汉英物理学词汇手册

编　　者：张军鹰

策划编辑：张春明　李法敏

责任编辑：贺　敏

序

改革开放30年来，我国的经济、科技、文化、教育、外贸和社会各项事业快速发展，取得了令人瞩目的成就，经济总量翻了两番之多，国内生产总值2007年已近25万亿元人民币，人均国民收入达到了2 000多美元，提前实现了小康目标；科技水平不断提高，高新技术快速发展，缩短了与世界先进国家的差距；文化、教育事业迅猛发展，不断满足和丰富着人民日益增长的文化精神需要；社会服务和保障体系不断完善，使中华民族和社会更加和谐。我国在国际事务中的作用日益凸显，与世界各国的政治、经济、科技、文化、教育、外交、军事等方面的交往日益频繁。成功加入世界贸易组织，成功申办和举办2008年北京奥运会和残奥会，成功申办上海世博会等一系列重大外交活动，塑造了中国在世界上的良好形象，更是促进了我国与世界各国的交往、交流和交融。经济全球化、科技一体化、文化多元化、信息网络化的发展趋势，使人们的生活正发生着深刻的变化。

新的学科、新的术语、新的词汇不断诞生和涌现。我国数亿不同层次的英语学习者和使用者，在学习或工作中不断遇到新的词语、新的术语。大部分的英语学习者和使用者能够比较顺利地用普通英语进行交流和交际，而一旦遇到专业领域的词汇或术语往往会陷入困境，有时可能了解某词语的一般意义，但不知道它在某些专业领域指的是什么。这种情况，在日常交往中，或在专业性较强的国际会议中屡见不鲜，常常需要英语专家和专业方面的人士通力合作，才能解疑释惑。若能编纂出版一套英汉、汉英的百科词汇手册，则将大大有益于英语学习者和使用者，能够为他们的英语学习和使用助上一臂之力。笔者曾在上世纪90年代初随同中国教材出版项目考察团访美，任翻译，就是依靠了一本英汉汉英出版印刷词汇手册，比较顺利地完成了任务。同时，此类专业词汇手册还有助于学习者提高英语水平和能力，借助英语汲取科技知识和信息，扩大视野，不断丰富和提高专业知识和水平。有鉴于此，上海外语教育出版社（以下简称外教社）与牛津大学出版社合作，于本世纪初陆续出版了牛津百科分类词典英语版40余种。这批百科词典的出版深受专业人员、英语学习者和使用者的欢迎。同时，根据部分读者的要求，外教社经过充分调研和论证，并征得牛津大学出版社同意，从该系列词典中挑选出近十种，组织有关专业人员编译成英汉双解版；由于工作量巨大，亦仅将这部分词典的词目翻译成汉语。随着对外交流和交往的深入发展，不断有读者询问外教社是否出版有完整的百科方面的词汇手册或词典。为满足这一需要，外教社经过广泛的调研、需求分析和市场论证，组织编写了外教社英汉·汉英百科词汇手册系列图书，历经四年的努力，全国近百位编纂者的通力合作和辛勤劳动，终于迎来了第一批书稿。

本百科词汇手册系列有以下几个特点：

1. 按学科、专业和行业分册编写（以下统称专业），分类依据主要是国家标准学科分类、国家标准国民经济行业分类、企业经营行业分类及中国图书在版编目分

类,涵盖自然科学、技术、社会科学、人文科学、社会生活等80余个领域;

2. 各专业词汇手册包括英汉、汉英两部分,尽可能涵盖各专业最基本、最常用的词汇,每部分收词基本上控制在5 000至10 000条,版式和开本设计便于使用和携带;

3. 各分册按词汇的使用频率收列专业基本词汇,同时力求反映该专业的最新发展,只收专业性词汇和普通语文词汇的专业性义项。与本专业关系不大的其他专业词汇尽量不收。

本系列词汇手册可供英语学习者、使用者及相关专业人员了解和熟悉专业词汇,学习和丰富专业知识,提升专业视野和水平之参考,亦可作为翻译的参考工具书。由于学识和能力所限,新的词汇层出不穷,收词难免挂一漏万,谬误和缺陷在所难免,敬请广大读者惠予批评、指正。

庄智象

2008年10月

前 言

物理学是自然科学中最基础的学科之一，也是一门在科技发展中发挥着重大作用的带头学科。物理学的历史渊源最早可以追溯到古希腊时期，当时的哲学家为了认识自然规律、解释物质起源作出了最初的研究尝试。

到19世纪末，经典物理学的大厦已基本筑就，形成了力学、声学、热力学与统计物理、电磁学和光学等众多分支学科。进入20世纪，以相对论和量子力学为标志的近代物理学的建立，更是将物理学的研究领域拓展到了前所未有的广度。

物理学在发展过程中形成了独特的研究方法，既包括从实验出发、从特殊到一般的归纳方法，又包括以理论为主导、从一般到特殊的演绎方法。事实证明，物理学所揭示的客观规律及独有的思想方法对自然科学乃至社会科学都产生了深远的影响，呈现出应用性极强、交叉领域极广的特点。

本词汇手册不仅收录了经典物理学、近代物理学各分支学科的专业词汇，也收录了与物理学密切相关的数学、化学、天文学、信息科学、电子学、材料科学等学科的常用词汇，英汉、汉英双向收词共17 000余条。本书可作为本学科领域的词汇速查、对照工具，供高校理工科师生及相关科研人员在查阅、翻译专业文献时参考使用。

由于编者的学识浅薄，词汇的选编范围及译文难免有错误之处，恳请热心的读者指出并予斧正。

使用说明

1. 本手册分为英汉、汉英两部分。
2. 英汉部分条目按照字母顺序排列;汉英部分条目按照拼音音序排列,同音字按照笔画数排列。
3. 英汉部分以数字、特殊符号、希腊字母等非英语字母开头的条目放在英汉部分正文之后;汉英部分以数字、特殊符号、西文字母等非汉字开头的条目放在汉英部分正文之后。
4. 外国人名条目的词头与译文一般采用姓前名后的形式。例如:

   Chomsky, Avram Noam 乔姆斯基,A·N

   乔姆斯基,A·N Chomsky, Avram Noam
5. 为便于查阅,汉英部分的某些条目(多为自由组合)以汉语中心词或关键词立目,其他文字置于空心鱼尾号中,放在中心词或关键词之后。例如:

   与产品相关的行为变量 product-related behavioural descriptor

   立目时处理为:

   行为变量〖与产品相关的〗 product-related behavioural descriptor
6. 英汉部分概念相同或相近的汉语译文之间用逗号分隔,概念不同的汉语译文之间用分号分隔;汉英部分的英语译文一律用分号间隔。
7. 圆括号“( )”用于括注:① 缩写或全称形式;② 可省略的内容;③ 解释说明性文字。
8. 方括号“[ ]”用于括注可替换的文字。
9. 尖括号“〈 〉”用于括注学科标注。

英汉部分

Aa

A (ampere) 安培(电流强度单位)
abampere 磁安
abat-voix 吸声板
Abbe double-diffraction principle 阿贝双衍射原理
Abbe principle of image formation 阿贝成像原理
Abelian collection 〈数〉阿贝尔群,交换群
Abelian group 〈数〉阿贝尔群,交换群
aberrated lens system 有像差透镜系统
aberrated optics 有像差光学部件;有像差光学系统
aberration 像差;〈天文〉光行差
aberration correction 像差校正
aberration curve 像差曲线
aberration-free system 消像差系统,无像差系统
abnormal absorption 反常吸收
abnormal current 反常流
abnormal magnetic moment 反常磁矩
abohm 磁欧,绝对欧姆
above-critical state 超临界状态
above-thermal neutron 超热中子
abrasive 研磨料,磨蚀剂
abrupt heterojunction 突变异质结
abrupt junction 突变结
abscissa 〈数〉横坐标
absolute acceleration 绝对加速度
absolute activity 绝对活度
absolute black body 绝对黑体
absolute displacement 绝对位移
absolute error 绝对误差
absolute expansion 绝对膨胀
absolute humidity 绝对湿度
absolute motion 绝对运动
absolute permittivity 绝对电容率
absolute pitch 绝对音调
absolute reference frame 绝对参照系
absolute scale of temperature 绝对温标
absolute temperature 绝对温度
absolute temperature scale 绝对温标
absolute time 绝对时间
absolute value 绝对值
absolute velocity 绝对速度
absolute zero 绝对零度
absorptance 吸收比,吸收度
absorption 吸收
absorption coefficient 吸收系数
absorption-free material 无吸收材料

absorption peak 吸收峰
absorption spectrum 吸收光谱，吸收频谱
absorptive power 吸收本领，吸收能力
absorptivity 吸收比，吸收率，吸收系数
abundance 丰度；分布量
abvolt 磁伏，绝对伏特
AC; a. c. (**alternating current**) 交流（电）
a.c. bridge 交流电桥
accelerated motion 加速运动
accelerating reference frame 加速参照系
acceleration 加速度
acceleration of free fall 自由落体加速度
acceleration of gravity 重力加速度
accelerator 加速器
accelerometer 加速度计，加速度测量仪
acceptor 接受器；受主
acceptor density 受主浓度
acceptor (energy) level 受主能级
acceptor impurity 受主杂质
acceptor ionization energy 受主电离能
accommodation 调节；调焦
accretion disk 〈天文〉吸积盘
accumulator 蓄电池；蓄能器
accuracy 准确度
achromat 消色差透镜
achromatic aberration 消色差
achromatic doublet 消色差双合透镜
achromaticity 消色差性
achromatic lens 消色差透镜
achromatic triplet 消色差三合透镜
achromatism 消色差性
acid 〈化〉酸
acoustic 声的，声波的
acoustical conductance 声导
acoustical conductivity 声导率；传声性
acoustical holography 声全息照相；声全息
acoustical lens 声透镜
acoustical perception 听觉，声觉
acoustical reactance 声抗
acoustical resistance 声阻
acoustical spectrometer 声谱计
acoustical spectroscopy 声谱学
acoustical wave 声波
acoustic emission 声发射
acoustic impedance 声阻抗
acoustic phonon 声频声子
acoustics 声学
acoustic susceptance 声纳
acoustic thermometer 声学低温计
acoustic velocity 声速
acoustimeter 声强计，测声计
acoustoelectric effect 声电效应
acoustoelectronic device 声电装置，电声器件
acoustooptic effect 声光效应
acoustooptic modulation 声光调制
acoustooptics 声光学
acting force 作用力
actinicity 光化性，光化度
actinic radiation 光化辐射
actinium series 〈化〉锕系
actinoelectricity 光电效应
actinology 光化学；射线化学
actinometer 化学光度计，光能测定仪
action 作用；作用量
action at a distance 超距作用
action potential 动作电位
action spectrum 作用光谱
activated atom 激活原子，活化原子
activated carbon 活性炭
activated carrier 激活载流子；激活载体
activated state 激活态，活化态
activating agent 激活剂，活化剂
activation 激活，活化，激发

activation analysis 激活分析,活化分析
activation energy 激活能,活化能
activation potential 活化势
active area 有效面积
active cavity 有源腔,激活腔
active current 有功电流,有效电流
active device 有源器件
active dipole 辐射偶极子
active electron 激活电子
active element 有源元件,有功元件
active galaxy 〈天文〉活动星系
active imaging system 有源成像系统
active level 激活能级
active load 有源负载,有功负荷
active mass 有效质量
active medium 激活介质
active nucleus 放射性核;活性核
active optical fibre 激活光纤
active power 有功功率,有效功率
active region 活动区,激活区
active section 活性区
active transport 主动输运
active voltage 有功电压,有效电压
activity 活性,活度,放射性活度
activity decay 活性衰减
actual output 实际输出功率
actual value 真值,实际值,绝对值
adapter 适配器;转接器
adaptive optical system 自适应光学系统
adaptive optics 自适应光学;自适应光学系统
adatom 吸附原子,增原子
ADC (analogue-to-digital converter) 模数转换器
adder 〈计算机〉加法器
adder amplifier 加法放大器
additive colour 相加色
additive colour mixing 加法混色
additive process 加色法
additivity law 加性定律
addressable memory 可寻址存储器
addressable register 可寻址寄存器,可编址寄存器
address decoder 地址译码器
adhesion 附着力;附着,黏附
adiabat 绝热线
adiabatic approximation 绝热式近似,浸渐近似
adiabatic chart 绝热图
adiabatic compression 绝热压缩
adiabatic compression heating 绝热压缩加热
adiabatic condition 绝热条件,浸渐条件
adiabatic demagnetization 绝热退磁
adiabatic diagram 绝热图
adiabatic equation 绝热方程
adiabatic equation of ideal gas 理想气体的绝热方程
adiabatic expansion 绝热膨胀
adiabatic exponent 绝热指数
adiabatic Hall effect 绝热霍尔效应
adiabatic process 绝热过程,浸渐过程
adjustable bearing 可调轴承
adjustable capacitor 可变电容器,可调电容器
adjustment 调节,调整
Adler anomaly 阿德勒反常
admissible error 容许误差
admittance 导纳
admixture 混合;混合物
adsorbability 吸附能力
adsorbate 被吸附物
adsorbent 吸附剂
adsorption 吸附
adsorption site 吸附座
advanced gas-cooled reactor (AGR) 改进型气冷反应堆
advancing wave 前进波
advection 平流,对流
advective layer 平流层

aeolotropic crystal 各向异性晶体
aeolotropism 各向异性
aeolotropy 各向异性
aerial 天线
aerial array 天线阵
aerial coupling 天线耦合
aerial effect 天线效应
aerial gain 天线增益
aerial mast 天线杆
aerial tuning 天线调谐
aeroacoustics 空气声学,航空声学
aeroballistics 空气弹道学,航空弹道学
aerobiology 空气生物学,高空生物学
aerocamera 航空照相机
aerodynamic drag 气动阻力
aerodynamics 空气动力学
aeroelasticity 气动弹性;气动弹性力学
aeroembolism 气泡栓塞症,高空气栓病
aerofoil 翼型;机翼
aerogenerator 风力发电机
aerohydrodynamics 空气流体动力学
aerolite 陨石
aerology 高空气象学
aerometer 气体比重计
aeronautics 航空学
aerosol 浮质;气溶胶;气雾剂
aerospace 宇宙空间
AFM (atomic force microscope) 原子力显微镜
afocal imaging system 无焦成像系统
afterglow 余辉
afterheat 余热
afterimage 余像
aftertreatment 后处理
age-hardening 时效硬化
age of the earth 地球年龄
age of the universe 宇宙年龄
aging 老化,陈化
AGR (advanced gas-cooled reactor) 改进型气冷反应堆
AI (artificial intelligence) 人工智能
air 空气
air admission valve 进气阀
airborne contamination 空气污染,大气污染
airborne laser tracker (ALT) 机载激光跟踪器
airborne magnetometer 航空地磁仪,机载磁强计
airbrake 空气制动机,风闸,减速板
air conditioner 空气调节器,空调
air core 空气芯
air core coil 空心线圈
aircraft 航空器
air damping 空气阻尼
air dielectric 空气电介质
air dielectric capacitor 空气(电介质)电容器
air equivalent 空气当量
air escape valve 泄气阀
airflow 气流
airflow meter 空气流量计
airfoil 翼型;机翼
airfoil profile 翼型,翼剖面
airfoil section 翼剖面
air friction 空气阻力,空气摩擦
airglow 气辉,大气辉光
airglow emission 辉光放射,气辉发射
airglow layer 气辉层
air ionization 空气电离
airmeter 空气流速计,小风速计
air moisture 空气湿度
airphoto 航摄相片
air plug 气塞
air pollution 空气污染,大气污染
airspeed 空速
airstream 气流
air table 气垫桌,气动工作台
airtightness 气密性
Airy disk 艾里斑
albedo 反照率
algebra 〈数〉代数学

algebraic sum 代数和
algebra of logic 逻辑代数
algebra of matrices 矩阵代数
algorithm 算法
algorithm complexity 算法复杂性
aligner 准直器,校准器
alive circuit 有源电路,带电电路
alkali 〈化〉碱
alkali-antimonides 碱金属锑化物
alkaline metal 碱金属
alkalinity 〈化〉碱度,碱性
alkali spectrum 碱金属光谱
all-band aerial 全波天线
allotrope 同素异形体
allotropism 同素异形(现象)
allotropy 同素异形(现象)
allowable deviation 容许偏差,许用偏差
allowable error 容许误差
allowable load 容许负载,许可载荷
allowed band 容许带
allowed transition 容许跃迁;容许转变
alloy 合金
alloyed junction 合金结
alloyed junction transistor 合金结晶体管
alloy steel 合金钢
all parallel ADC 全并行模数转换器
all-pass filter 全通滤波器
all-pass optical filter 全通滤光片
all-serial ADC 全串行模数转换器
all-wave oscillator 全波带振荡器
all-wave receiver 全波带接收机
almost periodic oscillation 殆周期振荡
alnico 铝镍钴合金
alpax 硅铝(明)合金
alpha activity α放射性
alphabet laser 多掺激光器
alpha counter tube α粒子计数管
alpha decay α衰变
alpha-deuteron model α粒子-氘核模型
alpha-iron α铁
alpha particle α粒子
alpha radiation α辐射
alpha radioactivity α放射性
alpha ray α射线
alsimag 铝硅镁合金
ALT (airborne laser tracker) 机载激光跟踪器
alternate-line scanning 隔行扫描
alternating circuit 交流电路
alternating current (AC; a.c.) 交流(电)
alternating-current amplifier 交流放大器
alternating-current output 交流输出
alternating electric field 交变电场,交流电场
alternating electromotive force 交变电动势
alternating flux 交变通量,交变磁通量
alternating magnetic field 交变磁场
alternating voltage 交变电压,交流电压
alternative energy 替代能源
alternator 交流发电机
altimeter 测高仪,测高计,高度计
altitude 地平纬度;高度,海拔
ALU (arithmetic logic unit) 算术逻辑单元
alumel 镍铝硅锰(电偶)合金
AM (amplitude modulation) 振幅调制,调幅
amalgam 汞合金,汞齐
ambiplasma 双极性等离子体
ambipolar 二极的,双极性的
American National Standards Institute (ANSI) 美国国家标准协会
American Standard Code for Information Interchange (ASCII) 美国信息交换标准码

ammeter 安培计,电流表
ammonia clock 氨分子钟
amorphous 非晶的,无定形的
amorphous dielectrics 非晶介电体
amorphous matter 非晶体
amorphous solid 无定形体,非晶态固体
amorphous state 非晶态
amount of substance 物质总量,材料用量
ampere (A) 安培(电流强度单位)
ampere-hour 安(培小)时
ampere-meter 安培表,电流表
Ampère force 安培力
Ampère's law 安培定律
Ampère's rule 安培定则
ampere-turn 安(培)匝数,安匝
ampholyte ion 两性离子
amphoteric impurity 双性杂质
amplification by R-C coupling 阻容耦合放大
amplification coefficient 放大系数
amplification ratio 放大率
amplified spontaneous emission 放大自发发射
amplifier 放大器;扩音器
amplifier-inverter 倒相放大器
amplitude 振幅,峰值
amplitude detection 幅度检波
amplitude distortion 振幅畸变,波幅畸变
amplitude grating 振幅型光栅,幅光栅
amplitude modulated light 调幅光
amplitude modulation (AM) 振幅调制,调幅
a.m.u. (atomic mass unit) 原子质量单位
analogue adder 模拟加法器
analogue circuit 模拟电路
analogue signal 模拟信号
analogue-to-digital conversion 模数转换
analogue-to-digital converter (ADC) 模数转换器
analysis of regression 回归分析
analysis of variance 方差分析
analytical balance 分析天平
analytical geometry 解析几何学
analytical mechanics 分析力学
analytic function 解析函数
analyzer 检偏器;分析器;分析镜
anamorphic optical system 变形光学系统
anamorphose 图像变形,图像失真
anamorphote lens 像歪曲透镜,变形镜头
anastigmat 消像散透镜,消像差透镜
anastigmatic lens 去像散镜头
anchor ring 锚环
AND gate 与门
AND-NOT gate 与非门
AND-OR gate 与或门
AND-OR-INVERT gate 与或非门
anechoic chamber 消声室
anemometer 风力计,风速计
aneroid barometer 无液气压计,空盒气压表
angle mirror 角(反射)镜
angle modulation 角调制
angle of declination 偏角
angle of deviation 偏向角
angle of diffraction 衍射角
angle of dip 倾角,真倾角
angle of divergence 发散角
angle of incidence 入射角
angle of polarization 偏振角
angle of reflection 反射角
angle of refraction 折射角
angle of rotation 转动角,自转角
angle of total reflection 全反射角
angstrom 埃(长度单位)
angular acceleration 角加速度
angular aperture 角孔径,孔径角
angular dispersion 角色散
angular displacement 角位移

angular distance 角距
angular frequency 角频率,圆频率
angular magnification 角放大率
angular magnifying power 角放大率
angular momentum 动量矩,角动量
angular-momentum conservation 角动量守恒
angular-momentum conservation law 角动量守恒定律,动量矩守恒定律
angular motion 角运动
angular quantum number 角量子数
angular resolution 角分辨率
angular speed 角速度
angular velocity 角速度
anharmonic force 非谐力
anharmonicity 非(简)谐性
anharmonic oscillator 非谐振子;非谐振荡器
anharmonic vibration 非谐振动
anion 负离子,阴离子
anisotropic 各向异性的
anisotropic crystal 各向异性晶体
anisotropic medium 各向异性介质,各向异性媒质
anisotropism 各向异性
annealing 退火
annihilation 湮没
annihilation radiation 湮没辐射
annular eclipse 〈天文〉(日)环食
annulus 环,环形物;圆环域
anode 阳极,正极
anomalous absorption 反常吸收
anomalous diffusion 反常扩散
anomalous dispersion 反常色散
anomalous scattering 反常散射
anomalous transport 反常输运
anomalous Zeeman effect 反常塞曼效应
anomaly 异常,反常;近点角,近点距
ANSI (**American National Standards Institute**) 美国国家标准协会
anthropic principle 人存原理
antiatom 反原子
antibaryon 反重子
antiboson 反玻色子
antideuteron 反氘核
antidiffusion 反扩散
antielectron 正电子,正子
antifermion 反费米子
antiferroelectrics 反铁电体
antiferromagnetism 反铁磁性
antigalaxy 〈天文〉反星系
antikaon 反K介子
antilepton 反轻子
antilogarithm 〈数〉反对数
antimatter 反物质
antineutrino 反中微子
antineutron 反中子
antinode 波腹
antinucleus 反核
antiparallel spins 反平行自旋
antiparallel vectors 反平行向量
antiparticle 反粒子
antiphase 反相(位),逆相(位)
antiproton 反质子
antiquark 反夸克
antireflecting film 减反射膜,增透膜
antistatic material 防静电材料
antiworld 反(物质)世界
anyon 任意子
aperiodicity 非周期性,无周期性
aperture 孔;光圈;孔径
aperture angle 孔径角
aperture stop 孔(径光)阑
aperture synthesis 孔径综合
aphelion 〈天文〉远日点
aplanat 消球差透镜,齐明镜,不晕镜
aplanatic lens 消球差透镜,齐明透镜
apoapsis 〈天文〉远点,远拱点
apochromat 复消色差透镜
apochromatic lens 复消色差透镜
apocynthion 〈天文〉远月点

apogee 〈天文〉远地点
apolune 〈天文〉远月点
apparent depth 表观深度
apparent expansion 表观膨胀,视膨胀
apparent mass 表观质量
apparent stellar brightness 〈天文〉恒星视亮度
apparent stellar magnitude 〈天文〉视星等
Appleton layer F电离层,阿普尔顿层
applications software 〈计算机〉应用软件
applied mechanics 应用力学
applied optics 应用光学
applied physics 应用物理(学)
approximation technique 近似技术
aqueous 含水的
arc discharge 电弧放电,弧光放电
Archimedes principle 阿基米德原理
arc lamp 弧光灯
Argand diagram 阿根图
argon ion etching 氩离子刻蚀
argon ion laser 氩离子激光器
argument 辐角;自变数
arithmetic average 算术平均数,算术平均值
arithmetic logic unit (ALU) 算术逻辑单元
arithmetic mean 算术平均,算术平均数[值]
arithmetic progression 算术级数,等差级数
arithmetic series 算术级数,等差级数,等差数列
armature 电枢;衔铁
arm of couple 力偶臂
arm of force 力臂
arm of the bridge 〈电子〉桥臂
artificial intelligence (AI) 人工智能
artificial radioactivity 人工放射性
artificial satellite 人造卫星
artificial transmutation 人为嬗变
ascensional force 升力
ASCII (American Standard Code for Information Interchange) 美国信息交换标准码
associative law 结合律,缔合律
astatic galvanometer 无定向电流计
astatic magnetometer 无定向磁强计
asteroid 〈天文〉小行星
asteroid belt 〈天文〉小行星带
asthenosphere 软流层,软流圈
astigmatism 散光,像散性,像散现象
astrometry 天体测量学
astronautics 航天学,宇航学
astronomical optics 天文光学
astronomical telescope 〈天文〉天文望远镜
astronomical unit (AU) 〈天文〉天文单位
astronomy 天文学
astrophysics 天体物理学
astrospectroscopy 天体光谱学
asymmetrical fission 不对称裂变
asymmetric atom 不对称原子
asymptote 〈数〉渐近线
asymptotic freedom 渐近自由
atmolysis 分气
atmosphere 大气;大气压(压力单位)
atmospheric pressure 大气压
atom 原子
atomic absorption spectroscopy 原子吸收光谱学
atomic beam 原子束
atomic bomb 原子弹
atomic clock 原子钟
atomic collision 原子碰撞
atomic energy 原子能,核能
atomic force microscope (AFM) 原子力显微镜
atomic fusion 核聚变

atomic mass 原子质量
atomic mass unit (a.m.u.) 原子质量单位
atomic model 原子模型
atomic nuclear physics 原子核物理学
atomic nucleus 原子核
atomic number 〈化〉原子序数
atomic orbital 原子轨道函数
atomic physics 原子物理学
atomic pile 原子反应堆
atomic radiation 原子辐射
atomics 原子科学
atomic scattering factor 原子散射因子
atomic spectral line 原子谱线
atomic spectrum 原子光谱
atomic spin 原子自旋
atomic structure 原子结构
atomic theory 原子论
atomic volume 原子体积
atomic weight 原子量
attenuation 衰减,衰耗
attenuation coefficient 衰减系数
attitude angle 姿态角
attraction 吸引,吸引力
attractive force 引力,吸引力
attractor 吸引子
Atwood machine 阿特伍德机
AU (astronomical unit) 〈天文〉天文单位
audibility 可听度
audiofrequency 音频
audiofrequency region 声频范围
audiometer 听度计,听力计
audio oscillator 声频振荡器
Auger effect 俄歇效应
Auger electron 俄歇电子
aurora 极光
aurora australis 南极光
aurora borealis 北极光
austenite 奥氏体
autoclave 高压釜,加压釜
autocollimatic spectroscope 自准光谱仪
autocollimation 自准直
autodoping 自掺杂
autoradiography 自动射线照相术,放射自显影法
autotransformer 自耦变压器
autovibration 自激振动
avalanche 雪崩
avalanche breakdown 雪崩击穿
average 平均数;平均的
average error 平均误差
average speed 平均速率
avionics 航空电子学
Avogadro constant 阿伏伽德罗常量
Avogadro's law 阿伏伽德罗定律
avometer 万用电表,安伏欧计
axial aberration 轴向像差
axial acceleration 轴向加速度
axial force 轴向力
axial magnification 轴向放大率
axial symmetry 轴向对称
axial vector 轴向量,轴矢(量)
axion 轴子
axis 轴
axis of incidence 入射轴
axis of inertia 惯量轴,惯性轴
axis of rotation 旋转轴,转动轴
axis of symmetry 对称轴
azeotrope 〈化〉共沸混合物,恒沸物
azeotropic mixture 〈化〉共沸混合物,恒沸物
azimuth 方位,方位角;〈天文〉地平经度
azimuthal quantum number 角量子数
azimuth polarimeter 旋光仪,方位偏振仪

Bb

b (barn) 靶(恩)(面积单位)
Babbitt metal 巴比特合金
Babinet compensator 巴比涅补偿器
Babinet principle 巴比涅原理
back diffusion 反向扩散
back discharge 负效放电
back e. m. f. 反电动势
backfeed loop 反馈电路
background radiation 本底辐射，背景辐射
backing store 〈计算机〉后备存储器
backup 备用元件;〈计算机〉备份
backward diode 反向二极管
balance 平衡;天平,秤
balanced bridge 平衡电桥
balance equation 平衡方程
ballast resistor 镇流电阻器,稳流电阻器
ballistic curve 弹道;弹道曲线
ballistic galvanometer 冲击电流计
ballistic pendulum 冲击摆,弹道摆
ballistic rocket 弹道火箭
ballistics 弹道学
ball lightning 球状闪电
Balmer series 巴耳末(谱线)系
band 频带;波带;光谱带;光带;能带
band elimination filter 带阻滤波器
band filter 带通滤光片,带通滤波器
band gap 带隙
band narrowing 能带变窄
band spectrum 带状谱,带光谱
band suppression filter 带阻滤波器
band theory 能带论,能带说
band-to-band transition 带间跃迁
bandwidth 带宽
bar 巴(压强单位)
Bardeen-Shockley formula 巴丁-肖克莱公式
bare charge 裸电荷
bare wire 裸线
barion 重子,激子
Barkhausen effect 巴克豪森效应
barn (b) 靶(恩)(面积单位)
barograph (自动)气压计
barometer 气压计
barometric pressure 大气压
barrel distortion 桶形畸变,桶形失真,负畸变
barrier 势垒;阻挡层
barrier effect 势垒效应
barrier height 势垒高度
barrier penetration 势垒穿透
barycentre 重心,质心,引力中心
barye 微巴(压强单位)
baryon 重子
baryon number 重子数
base 基极;〈数〉基数
base doping 基区掺杂
base unit 基本单位
basic frequency 主频(率)
battery 电池(组);蓄电池
battery charger 电池充电器
baud (Bd) 波特(信号传输速率单位)
Bayes theorem 贝叶斯定理

b.c.c. (body-centred cubic) 体心立方体
BCS theory BCS 理论
Bd (baud) 波特(信号传输速率单位)
beam 梁;束
beam angle 光束孔径角,射束孔径角,束锥角
beam balance 杠杆式天平
beam expander 扩束器
beam hole 束孔
beam intensity 束流强度
beam radius (光)束半径
beam ratio 束强比,束比
beam spot (光)束斑,射束点
beam waist (光)束腰
beam waveguide 光束波导
beat frequency 拍频
beats 拍,差拍
Beckmann thermometer 贝克曼(水银)温度计
becquerel (Bq) 贝可勒尔,贝可(放射性活度单位)
bel 贝尔(音量比率单位)
bell metal 钟铜,钟青铜
Bell's theorem 贝尔定理
Bénard cell 贝纳胞
bending moment 挠矩,弯(曲)矩
bending strain 弯曲应变,挠胁变
bending strength 抗弯强度,抗挠强度
bending stress 弯曲应力,挠胁强
Bernoulli equation 伯努利方程
Bernoulli theorem 伯努利定理
Berthollide compound 变比化合物,贝陀立式化合物
Bessel function 贝塞尔函数
Bessemer process 贝塞麦炼钢法,酸性转炉炼钢法
beta decay β衰变
beta-iron β铁
beta particle β粒子
beta radiation β辐射
beta radioactivity β放射性
beta ray β射线
betatron 电子感应加速器
BeV (billion-electronvolt) 千兆电子伏特,十亿电子伏特
Bevatron 高能质子同步稳相加速器
biaxial crystal 双轴晶体
biconcave lens 双凹透镜
biconvex lens 双凸透镜
bicrystal 孪晶体,双晶体
bifilar pendulum 双线摆
bifocal lens 双焦透镜
big-bang cosmology 大爆炸宇宙论
big-bang theory 大爆炸论
bilateral constraint 双侧约束,双边约束
bilepton 双轻子
bilinear coupling 双线性耦合
billion 吉(咖),千兆,十亿
billion-electronvolt (BeV) 千兆电子伏特,十亿电子伏特
bimetallic strip 双金属片
bimetallic thermometer 双金属温度计
bimolecular recombination 双分子复合
bimorph cell 双压电晶片元件
binary collision 二体碰撞,两体碰撞
binary digit 二进制数字
binary fission 二分裂变
binary hologram 二元全息图,二进制全息图
binary notation 二进制记数法
binary stars 双星
binding energy 结合能
binocular field glasses 双目望远镜,双筒望远镜
binocular microscope 双目显微镜,双筒显微镜
binoculars 双目镜,双筒镜
binocular telescope 双目望远镜,双筒望远镜
binocular vision 双目视觉
binomial expansion 二项展开式
binomial theorem 二项式定理
bioenergetics 生物能学

biological microscope 生物显微镜
bioluminescence 生物发光,生物荧光
biomechanics 生物力学
biophysics 生物物理学
biotechnology 生物工艺学
Biot-Savart law 毕奥-萨伐尔定律
bipolaron 双极化子
biprism 双棱镜
birefringence 双折射
bistability 双稳性
bistable circuit 双稳态电路
bit 比特
Bitter powder pattern 比特粉纹图样
black-and-white group 黑白群
black body 黑体
black-body radiation 黑体辐射
black hole 黑洞
blanket gas 保护气体
blast furnace 鼓风炉
blast wave 冲击波,爆炸波
blazed grating 闪耀光栅
blind spot 盲点
Bloch's theorem 布洛赫定理
Bloch wave function 布洛赫波函数
block and tackle 滑轮组,滑车组
blocking effect 阻塞效应,遮断效应
block prism 方块棱镜,立方棱镜
blooming 敷霜,涂膜
blue phase 蓝相
blue shift 蓝(色偏)移
blurred image 模糊图像
body-centred cubic (b.c.c.) 体心立方体
body-centred lattice 体心格,体心点阵
Bohr atom model 玻尔原子模型
Bohr frequency condition 玻尔频率条件
Bohr magneton 玻尔磁子
Bohr radius 玻尔半径
Bohr theory 玻尔理论,玻尔学说
boiling 沸腾
boiling point (b.p.) 沸点
boiling point diagram 沸点图
boiling-water reactor (BWR) 沸水反应堆
bolometer 辐射热计
Boltzmann constant 玻尔兹曼常量
Boltzmann distribution law 玻尔兹曼分布律
Boltzmann equation 玻尔兹曼方程
Boltzmann factor 玻尔兹曼因子
bombarding particle 轰击粒子,施轰粒子
bomb calorimeter 爆炸量热器
bond distance 键长,键距
bond energy 键能
bonding 成键,键合
bonding electron 成键电子
Boolean algebra 布尔代数
Born-Green equation 玻恩-格林方程
Born-Haber cycle 玻恩-哈伯循环
Born-Mayer potential 玻恩-迈耶势
Born-Oppenheimer approximation 玻恩-奥本海默近似
boron chamber 硼电离室
boron counter tube 硼计数管
Bose-Einstein condensation 玻色-爱因斯坦凝聚
Bose-Einstein statistics 玻色-爱因斯坦统计法
boson 玻色子
bosonic string 玻色(子)弦
bottom quark b夸克,底夸克
boundary condition 边界条件
boundary layer 边界层,界面层
boundary relation 边值关系
boundary scattering 边界散射
boundary-value problem 边值问题
bound charge 束缚电荷
bound electron 束缚电子

bound state 束缚态
Bourdon gauge 布尔登压力计
bow wave 头波,舷波,涡流
Boyle's law 玻意耳定律
b.p.(boiling point) 沸点
b-quark b 夸克,底夸克
Bq(becquerel) 贝可勒尔,贝可(放射性活度单位)
Brackett series 布拉开系
Bragg diffraction 布拉格衍射
Bragg's law 布拉格定律
Bragg vector 布拉格矢量
brake 制动闸,轫
brake pedal 刹车踏板
braking distance 制动距离
branch 支路
brass 黄铜
Bravais group 布拉维群
Bravais lattice 布拉维格
breadboard 模拟板,面包板
breakdown 击穿
breakdown field strength 击穿场强
breakdown voltage 击穿电压
breaker 断路器
breaking strength 断裂强度,抗断强度
breaking stress 致断胁强,致断应力
breeder reactor 增殖反应堆
Bremsstrahlung 轫致辐射
Brewster angle 布儒斯特角
Brewster's law 布儒斯特定律
Brewster window 布儒斯特窗
bridge bond 桥键
bridge circuit 桥式电路,桥接电路
bridge rectifier 桥式整流器
bridging atom 桥接原子
bright field 明(视)场,亮视场
bright line 明线
brightness 亮度
brilliance 耀度
Brillouin function 布里渊函数
Brillouin scattering 布里渊散射
Brillouin zone 布里渊区
Brinell hardness 〈冶金〉布氏硬度,压球硬度
Britannia metal 大不列颠合金
British system of units 英国单位制
British thermal unit(Btu) 英国热单位
brittle material 脆性材料
broadband 宽(谱)带
broadcast 广播
broken symmetry 破坏对称性
bronze 青铜
Brownian motion 布朗运动
brush 电刷
brush discharge 刷形放电,电刷放电
Btu(British thermal unit) 英国热单位
bubble chamber 气泡室,泡沫室
buffer circuit 缓冲电路
buffer register 缓冲寄存器
built-in field 内建场
bulk modulus 体积(弹性)模量,本体模量
bumping 爆沸,崩沸
bunch compression 群聚压缩
Bunsen burner 本生灯
Bunsen cell 本生电池
buoyancy 浮力
buoyancy force 浮力
burnishing 抛光
bus 总线,母线
butterfly effect 蝴蝶效应
BWR(boiling-water reactor) 沸水反应堆
by-pass 旁路
byte 〈计算机〉字节

Cc

C (coulomb) 库仑(电量单位)
cable 电缆
CAD (computer-aided design) 计算机辅助设计
cadmium cell 镉电池
caesium clock 铯(原子)钟
calcite 方解石
calculus 〈数〉微积分;演算
caloric theory of heat 热质说
Calorie 大卡,千卡(热量单位)
calorie 卡路里(热量单位)
calorific value 热值
calorimeter 量热器,热量计,量能器
calx 金属灰,生石灰,矿灰
camera 照相机;摄影机,摄像机
Canada balsam 加拿大胶
canal ray 阳极射线,(电子管的)极隧射线
candela (cd) 坎(德拉)(发光强度单位)
candle power 烛光(照明单位)
canonical coordinates 正则坐标
canonical distribution 正则分布
canonical equation 正则方程
canonical transformation 正则变换
capacitance 电容
capacitive reactance 容抗
capacitor 电容器
capacitor microphone 静电传声器
capacity 容量;电容
capillarity 毛细现象,毛细作用
capillary action 毛细作用
capillary bore 毛细管孔
capillary tube 毛细管
capillary wave 表面张力波,界面波
capture 俘获
capture cross-section 俘获截面
carat 克拉(宝石重量单位);开金
carbon cycle 〈生态〉碳循环
carbon dating 碳年代法
carbon dioxide 二氧化碳
carbon dioxide laser 二氧化碳激光器
carbon fibres 碳纤维,碳丝
cardinal plane 基面
cardinal point 基点
Carnot cycle 卡诺循环
Carnot efficiency 卡诺效率
Carnot theorem 卡诺定理
carrier 载波;载流子;载体
carrier concentration 载流子浓度
carrier density 载流子密度
carrier gas 运载气体
carrier injection 载流子注入
carrier lifetime 载流子寿命
carrier mobility 载流子迁移率
carrier recombination 载流子复合
carrier rocket 运载火箭
carrier transport 载流子输运
carrier wave 载波
carrier-wave frequency 载波频率
Cartesian coordinates 笛卡儿坐标,直角坐标
cascade cooling 级联冷却
cascade decay 级联衰变
cascade emission 级联发射
cascade laser 级联激光器,多级激光器
cascade process 级联过程

cascade radiation 级联辐射,级联放射
cascade transformer 级联变压器;串级变压器
cascade transition 级联跃迁
case hardening 表面硬化
Casimir effect 卡西米尔效应
Cassegrain telescope 卡塞格林望远镜
cast iron 铸铁,生铁
CAT (computerized axial tomography) 计算机轴向层析成像
catadioptric system 反射折射光学系统,反折系统
catadioptric telescope 反射折射望远镜
catalytic converter 催化转化器
cataphoresis 电透法,电泳
catastrophe 突变,灾变
catastrophe theory 突变理论
catastrophic laser-damage threshold 灾变性激光损伤阈
catenary 〈数〉悬链线
cathetometer 高度计,高差计,测高计
cathode 阴极,负极
cathode ray 阴极射线
cathode-ray oscilloscope (CRO) 阴极射线示波器
cathode-ray tube (CRT) 阴极射线管
cathodic protection 阴极保护,阴极防蚀
cathodoluminescence 阴极射线发光
cation 阳离子,正离子
catoptric imaging 反射成像
catoptrics 反射光学
Cauchy dispersion formula 柯西色散公式
causality 因果性,因果关系;因果律
caustic 焦散的;腐蚀性的
caustics 焦散线
cavitation 空化,汽蚀,空穴现象
caviton 腔子
cavity 腔,空穴
cavity-dumped laser 倾腔激光器
cavity mode 空腔谐振模,腔模
cavity oscillation 腔振荡
cavity resonator 谐振腔,腔体谐振器
cc (cubic centimetre) 立方厘米(体积单位)
CCD (charge coupled device) 电荷耦合器件,电荷耦合装置
CD (compact disk) 光盘
cd (candela) 坎(德拉)(发光强度单位)
CD-I (CD interactive) 交互式 CD
CD-ROM (CD read-only memory) 光盘驱动器
CdS laser 硫化镉激光器
CDW (charge density wave) 电荷密度波
celescope 天体望远镜
celestial body 天体
celestial equator 〈天文〉天赤道
celestial mechanics 天体力学
celestial photography 天体照相术
celestial sphere 〈天文〉天球
cell 电池;晶胞;单元;细胞
cell spin 元胞自旋
cellular method 元胞法,网格法
Celsius (temperature) scale 摄氏温标
Celsius thermometer 摄氏温度计
Celsius thermometric scale 摄氏温标
cementation 粘固,胶合;〈冶金〉渗碳
cemented doublet 胶合双透镜,双胶透镜
cementite 渗碳体,碳化铁
centigrade scale 百分标,百分刻度,摄氏温标
centimetre (cm) 厘米(长度单位)
central collision 对心碰撞
central field 有心力场,辏力场

central force 有心力,辏力
central fringe 中央条纹
central maximum 中央极大
central processing unit (CPU) 〈计算机〉中央处理单元
central symmetry 中心对称,点对称
centred system 合轴组
centre of curvature 曲率中心
centre of force 力心
centre of gravity 重心
centre of mass 质心
centre-of-mass system (CMS) 质心系,质心参考系
centre of percussion 撞击中心,冲击中心
centrifugal acceleration 离心加速度
centrifugal barrier 离心势垒
centrifugal force 离心力
centrifugal pump 离心泵,离心抽机
centrifugation 离心,离心分离
centrifuge 离心机,离心分离机
centripetal acceleration 向心加速度
centripetal force 向心力
centroid 质心;形心
Cerenkov counter 切连科夫计数器,切连科夫探测器
Cerenkov detector 切连科夫计数器,切连科夫探测器
Cerenkov radiation 切连科夫辐射
cermet 金属陶瓷,含陶合金
CERN (Conseil Européen pour la Recherche Nucléaire) 欧洲粒子物理研究所
c.g.s. system c.g.s.单位制,绝对单位制,厘米·克·秒制
c.g.s. units 厘米·克·秒单位
chain reaction 链式反应
chain structure 链状结构
Chandler number 钱德勒数
Chandrasekhar limit 钱德拉塞卡极限
change of phase 相变
change of state 物态变化
channel 道;信道;(束流)孔道;通道
channel capacity 信道容量,通路容量
channeling effect 沟道效应
chaos 混沌
chaotic field 混沌场
chaotic laser light 混沌激光
chaotic measure 混沌测度
chaotic motion 混沌运动
characteristic 特征数;(对数的)首数
characteristic curve 特性曲线
characteristic equation 特征方程,久期方程
characteristic frequency 特征频率
characteristic function 特性函数
characteristic spectrum 特征光谱,标志光谱
characteristic state 特征状态,本征态
characteristic X-ray 特征X射线
charge 电荷;充电
charge asymmetry 电荷不对称性
charge carrier 载流子,载荷子
charge compensation 电荷补偿
charge conjugation 正反共轭,电荷共轭
charge coupled device (CCD) 电荷耦合器件,电荷耦合装置
charged body 带电体
charged current 电性流
charge density 电荷密度
charge density wave (CDW) 电荷密度波
charged particle 带电粒子
charged particle beam fusion 带电粒子束聚变
charged particle reaction 带电粒子反应
charge exchange (等离子)电荷交换
charge independence 电荷无关性

charge injection 电荷注入
charge injection device（CID）电荷注入器件
charge invariance 电荷不变性
chargeless particle 中性粒子，不带电粒子
charge-mass ratio 荷质比
charge operator 电荷算符
charge radius 电荷半径
charge symmetry 电荷对称性
charge transfer 电荷转移
charge transfer device 电荷转移器件
charge transfer process 电荷转移过程
charging by friction 摩擦起电
charging time constant 荷电时间常量，充电时间常数
Charles' law 查理定律
charm 粲，粲数
chart of nuclides 核素图
chemical affinity 化学亲和势；化学亲和性
chemical bond 化学键
chemical bonding 化学成键
chemical cell 化学电池
chemical energy 化学能
chemical physics 化学物理学
chemical potential 化学势
chemical vapour deposition（CVD）化学汽相沉积
chemi-ionization 化学电离
chemiluminescence 化合光，化学发光，化学荧光
chemisorption 化学吸附
Chinese Physical Society 中国物理学会
chip 芯片，晶片；石屑
chiral charge 手征荷
chiral compound 手征化合物
chiral current 手征流
chirality 手征性，手性
Chiron 〈天文〉喀戎星（小行星2060号）
Chladni figures 克拉尼图形
choke 扼流圈
choking coil 扼流线圈
cholesteric liquid crystal 胆甾型液晶
cholesteric phase 螺状相，胆甾相，胆甾醇相
chromatic aberration 色像差
chromaticity 色品，色度，染色性
chromaticity diagram 色品图，色度图
chromatic polarization 色偏振
chromatism 色差，变色
chromatism of magnification 放大率色差
chromatism of position 位置色差
chromatogram 色谱图，色层分离谱
chromatograph 色谱仪
chromatography 色谱法，色层分析法
chromatometer 色度计，色觉计
chrominance 色度
chromium steel 铬钢
chromometer 色度计，比色计
chromosphere 〈天文〉色球，色球层
CID（charge injection device）电荷注入器件
ciliary body 睫状体
circle 循环，周期；圆
circuit 电路，线路，回路
circuital theorem of electrostatic field 静电场环路定理
circuit breaker 断路器，断路开关
circuit diagram 电路图，线路图
circular birefringence 圆双折射
circular disk diffraction 圆盘衍射
circular frequency 圆频率
circular hole diffraction 圆孔衍射
circular measure 弧度，弧度法
circular motion 圆周运动
circular polarization 圆偏振
circulation 环流，环量
cistern barometer 槽式气压计，液槽气压计，杯式（水银）气压计

CL (confidence level) 置信水平,置信度
cladding 包覆涂层,覆层,镀层
Clapeyron-Clausius equation 克拉珀龙-克劳修斯方程
Clark cell 克拉克电池
classical coherence 经典相干性
classical degree of freedom 经典自由度
classical electrodynamics 经典电动力学
classical electron radius 经典电子半径
classical electron theory of metal 经典金属电子论
classical field theory 经典场论
classical mechanics 经典力学,牛顿力学
classical optics 经典光学
classical physics 经典物理学
classical spin model 经典自旋模型
classical statistics 经典统计法
Claude cycle 克洛德循环
clausius 克劳(熵的单位)
Clausius-Mossotti equation 克劳修斯-莫索提方程
cleaning 净化,清洁处理,清洗
clean surface 洁净表面
cleavage 解理,劈理
clinical thermometer 体温计
closed circuit 闭(合电)路
closed resonator 闭(共振)腔,闭合共振器
closed system 闭系
closed universe 闭宇宙
close packing 密堆积
cloud chamber 云室
clump 凝团
cluster 集团,簇,群集
cm (centimetre) 厘米(长度单位)
CMOS (complementary metal-oxide semiconductor) 互补型金属氧化物半导体
CMS (centre-of-mass system) 质心系,质心参考系
co-adsorption 共吸附
coagulation 凝结
coal 煤
coarse adjustment 粗调(节)
coarse-grained density 粗粒密度
coaxial cable 同轴线,同轴电缆
coaxiality 共轴性
cobalt steel 钴钢
COBE (cosmic background explorer) 宇宙背景探测器
Cockcroft-Walton accelerator 高压倍加器,科克罗夫特-瓦尔顿加速器
coefficient 系数
coefficient of diffusion 扩散系数
coefficient of expansion 膨胀系数
coefficient of friction 摩擦系数
coefficient of restitution 回弹系数,恢复系数
coefficient of sliding friction 滑动摩擦系数
coefficient of static friction 静摩擦系数
coefficient of stiffness 劲度系数,倔强系数
coefficient of viscosity 黏性系数,黏滞系数
coelostat 〈天文〉定天镜
coercive electric field 矫顽电场
coercive force 抗磁力,矫顽力
coercive stress 矫顽应力
coercivity 矫顽力,矫顽性
coherence 相干性
coherence condition 相干条件
coherence effect 相干效应
coherence imaging 相干成像
coherence length 相干长度
coherence time 相干时间
coherent light 相干光
coherent radiation 相干辐射
coherent scattering 相干散射
coherent source 相干光源
coherent state 相干态
coherent units 一贯制单位
coherent wave 相干波

cohesion 内聚力
cohesive energy 内聚能
cohesive force 内聚力;黏附力
coil 线圈;盘管
coincidence circuit 符合电路
coincidence counting 符合计数
coincidence method 符合法
coincidence prism 叠像棱镜
coincidence range finder 叠像测距仪
coincidence rate 符合率
coincidence spectrum 符合光谱
coincidence structure 符合结构
cold emission 冷发射
cold fusion 冷聚变
cold light 冷光
cold neutron 冷中子
cold plasma 冷等离子体
cold trap 冷阱,冷槽
collapsar 〈天文〉坍缩星
collapse 坍缩
collapse of shell 壳层坍缩
collapse-revivals 坍(缩恢)复现象
collective beating 集体拍
collective excitation 集体激发
collective frequency 集体频率
collective oscillation 集体振荡
collective rotation 集体转动
collective vibration 集体振动
collector 集电极;集光器
collider 对撞机
colliding beam 对撞束
colligative properties 依数性
collimated beam 准直光束
collimation 准直
collimator 平行光管;准直管;准直器
collinear transformation 直射变换,共线变换
collineation 直射变换,共线变换
collision 碰撞
collisional heating 碰撞加热
collisionally conserved quantity 碰撞守恒量
collision broadening 碰撞(谱线)增宽,碰撞展宽
collision cross-section 碰撞截面
collision density 碰撞密度
collision electrification 碰撞起电
collision frequency 碰撞频率
collisionless damping 无碰撞阻尼
collisionless instability 无碰撞不稳定性
collision process 碰撞过程
collision relaxation 碰撞弛豫
collision strength 碰撞强度
collision time 碰撞时间
colloid 胶体,胶质
cologarithm 〈数〉余对数
colorimeter 色度计,色量计
colorimetric pyrometer 比色高温计
colorimetry 色度学,比色法
colour 颜色
colour centre laser 色心激光器
colour charge 色荷
colour code 色标,色码
colour disk 色盘
coloured light 色光
coloured particle 色粒子
coloured quark 色夸克
colour filter 滤色器,滤色片
colour holography 彩色全息术
colour photography 彩色照相术
colour television 彩色电视
colour temperature 色温
colour vision 色视觉
coma 彗形像差;〈天文〉彗发
comatic aberration 彗形像差
combination 化合物;〈数〉组合
combination of lenses 透镜组
combination principle 组合原理,并合原理
combined parity 组合宇称
combustion 燃烧
comet 〈天文〉彗星
common logarithm 〈数〉常用对数
communication satellite 通信卫星

commutation relation 对易关系
commutative law 〈数〉交换律
commutator 对易式;换向器,转接器
compact disk（CD）光盘
compact galaxy 〈天文〉致密星系
compact object 〈天文〉致密天体
companion galaxy 〈天文〉伴星系
companion star 〈天文〉伴星
comparative half-life 比较半衰期
comparison spectrum 比较光谱
compass 罗盘,指南针;圆规
compensating eyepiece （色差)补偿目镜
compensating plate 补偿板
compiler 编译程序
complementarity 互补性
complementary colour 补色,互补色
complementary metal-oxide semiconductor（CMOS）互补型金属氧化物半导体
complementary principle 互补原理,并协原理
completely coherent light 完全相干光
completely inelastic collision 完全非弹性碰撞
complex amplitude 复振幅
complex conjugate 复共轭
complex degree of coherence 复相干度
complex filter 复合滤波器
complex impedance 复阻抗
complex index of refraction 复折射率
complex number 〈数〉复数
component 分量,组分;部件
component force 分力
component vectors 分矢量
component velocity 分速度
composite particle 复合粒子
composition of forces 力的合成,合力
composition of velocities 速度合成
compound 〈化〉化合物;混合物
compound lens 复合透镜
compound microscope 复显微镜
compound nucleus 复合核
compound-nucleus decay 复合核衰变
compound-nucleus theory 复合核理论
compound pendulum 复摆
compound system 复合系统
compressibility 压缩率;可压缩性
compression 压缩
compressional wave 压缩波
compression ratio 压缩率,压缩比
Compton effect 康普顿效应
Compton electron 康普顿电子
Compton scattering 康普顿散射
Compton wavelength 康普顿波长
computational physics 计算物理学
computer 计算机
computer-aided design（CAD）计算机辅助设计
computerized axial tomography（CAT）计算机轴向层析成像
computerized tomography（CT）计算机层析成像
concave 凹形的,凹面的
concave-convex lens 凹凸透镜
concave grating 凹面光栅
concave lens 凹透镜
concave mirror 凹面镜
concentration cell 浓差电池
concentric beam 同心光束
concentric lens 同心透镜
concentric resonator 共心(共振)腔
conchoidal fracture 贝壳状断裂,贝壳状裂痕
concurrent force 共点力
condensation 凝结,冷凝
condensation pump 冷凝泵,冷凝抽机

condensed matter 凝聚体
condensed-matter physics 凝聚体物理学,凝聚态物理学
condenser 电容器;冷凝器;聚光器
condenser microphone 静电传声器
condensing 凝结,冷凝
conductance 电导,导率,电导系数
conducting medium 导电介质
conduction 传导
conduction band 导带
conduction current 传导电流
conductive fibre 导电纤维
conductivity 电导率,导电性,传导率
conductivity water 校准电导水
conductor 导体
cone 圆锥体,锥形物;〈光〉(视网膜)锥体
cone effect 圆锥效应
cone emission 圆锥发射
confidence interval 置信区间
confidence level (CL) 置信水平,置信度
configuration 位形;组态,构造
configuration space 位形空间
confinement time 约束时间
confocal resonator 共焦(共振)腔
conformal transformation 共形变换,保角变换
conformational analysis 构形分析,构象分析
conic 〈数〉二次曲线
conjugate defect 共轭缺陷
conjugate field 共轭场
conjugate image 共轭像
conjugate planes 共轭面
conjugate points 共轭点
conjugate rays 共轭光线
conjugate wave 共轭波
conjugation operator 共轭算子
conjunction 〈天文〉合
connected bond 连通键
connection in parallel 并联
connection in series 串联
Conseil Européen pour la Recherche Nucléaire (CERN) 欧洲粒子物理研究所
conservation law 守恒律
conservation of angular momentum 角动量守恒
conservation of charge 电荷守恒
conservation of energy 能量守恒,能量不灭
conservation of mass 质量守恒,质量不灭
conservation of mechanical energy 机械能守恒
conservation of momentum 动量守恒
conservative field 保守场
conservative force 保守力
conservative system 保守系,守恒系
consonance 共鸣,谐和
constancy of light velocity 光速不变性
constant 常数,常量
constant acceleration 等加速度
constantan 康铜(铜镍合金)
constant-boiling mixture 〈化〉共沸混合物,恒沸物
constant current source 恒流源
constant force 恒力
constant of integration 〈数〉积分常数
constant speed 等速
constant voltage source 恒压源
constitutive equation 本构方程
constrained motion 约束运动
constraining force 约束力
constraint 约束,制约
constraint of position 位置约束
constraint of velocity 速度约束
constructive interference 相长干涉
contact force 接触力,触点(压)力
contact potential difference 接触电势差

containment shell (反应堆)防事故外壳
continuity equation 连续性方程
continuous eigenvalue 连续本征值
continuous function 〈数〉连续函数
continuous laser 连续激光
continuous medium 连续介质
continuous phase 连续相
continuous phase transition 连续相变
continuous spectrum 连续谱
continuous wave 连续波;等幅波
continuous wave laser 连续波激光器
continuum 连续区;连续统一体,连续统;连续介质
contrast 衬比度,反差
control grid 控制栅极
control parameter 控制参量
control rod 控制棒,控制杆
control unit (CU) 控制部件,控制器
convected acceleration 牵连加速度
convected inertial force 牵连惯性力
convected motion 牵连运动
convected velocity 牵连速度
convection 运流,对流
convection current 运流电流,对流电流
conventional current 常规电流
convergent lens 会聚透镜
convergent mirror 会聚镜
convergent wave 会聚波
conversion efficiency 转换效率
conversion electron 转换电子
conversion factor 换算因子,变换因数,转换因子;变流因数
converter 变流机,变流器;变频管,变频器
converter reactor 转换反应堆
convex 凸面的
convex lens 凸透镜
convex mirror 凸面镜
convey electron 运送电子
coolant 冷冻剂,冷却剂,散热剂
cooperative emission 合作发射
cooperative luminescence 合作发光
cooperative phenomenon 合作现象,协作现象
Cooper instability 库珀不稳定性
Cooper minimum 库珀最小
Cooper pairs 库珀对
coordinate 坐标
coordinate geometry 坐标几何
coordinate system 坐标系
coordination number 配位数
Copernican system 哥白尼体系
cophasal surface 等相面
coplanar force 共面力
core 核心;(线圈)磁芯
core loss 磁芯损耗
core polarization 芯极化,核心极化
Coriolis acceleration 科里奥利加速度
Coriolis force 科里奥利力
corkscrew rule 螺旋法则
corner cube 隅角棱镜,三面直角棱镜
corona 电晕,晕环;〈天文〉日冕
corona discharge 电晕放电
corpuscular property 粒子性,微粒性
corpuscular theory 微粒说
correcting field 校正场
correcting lens 校正透镜
correction 修正,校正
correction plate 校正板
correlation coefficient 关联系数,相关系数
correlation function 关联函数,相关函数
correlation spectrum 关联谱
correspondence principle 对应原理
corresponding state 对应态
corrosion 腐蚀

cosine rule 〈数〉余弦定律
cosmic abundance 宇宙丰度
cosmic background explorer (COBE) 宇宙背景探测器
cosmic background radiation 宇宙背景辐射
cosmic censorship hypothesis 宇宙监督假设
cosmic dust 宇宙尘
cosmic neutrino 宇宙中微子
cosmic noise 宇宙噪声
cosmic radiation 宇宙辐射
cosmic ray 宇宙线,宇宙光
cosmological constant 宇宙常数
cosmological principle 宇宙学原理
cosmological redshift 宇宙学红移
cosmology 宇宙学,宇宙论
cosmophysics 宇宙物理学
cosmos 宇宙
Cottrell electrostatic precipitator 科特雷尔静电除尘器
coudé telescope 折轴望远镜
coulomb (C) 库仑(电量单位)
Coulomb barrier 库仑势垒
Coulomb collision 库仑碰撞
Coulomb constant 库仑常数
Coulomb energy 库仑能
Coulomb excitation 库仑激发
Coulomb field 库仑场
Coulomb force 库仑力
Coulomb gauge 库仑规范
Coulomb's law 库仑定律
coulometer 电量计,库仑计
counter 计数器
counter e.m.f. 反电动势
counting rate 计数率
couple 力偶
coupled circuit 耦合电路
coupled pendulums 耦合摆
coupled resonators 耦合(共振)腔
coupling 耦合;结合
coupling coefficient 耦合系数
coupling constant 耦合常数
coupling scheme 耦合方式
covalent bond 共价键
covalent bonding 共价键合
covalent crystal 共价晶体
covalent graph 共价图
C-parity C宇称,正反共轭宇称
CP invariance CP不变性,电荷宇称守恒
CP parity CP宇称
CPT theorem CPT定理
CPU (central processing unit) 〈计算机〉中央处理单元
CP violation CP破坏,电荷宇称不守恒
c-quark c夸克,粲夸克
crack propagation 裂纹扩展
cranked eyepiece 转向目镜
crankshaft 曲轴
creep 蠕变
creep film effect 爬膜效应
crest 波峰
critical angle 临界角
critical damping 临界阻尼
critical length 临界长度
critical mass 临界质量
critical parameter 临界参量
critical phenomenon 临界现象
critical point 临界点
critical pressure 临界压强,临界压力
critical reaction 临界反应,临界回授
critical state 临界态
critical surface 临界面
critical temperature 临界温度
critical velocity 临界速度
critical volume 临界体积
CRO (cathode-ray oscilloscope) 阴极射线示波器
crossed polarizers 正交偏振器
cross prisms 正交棱镜
cross product 向量积,叉积
cross-section (横)截面,横断面
CRT (cathode-ray tube) 阴极射线管
crucible 坩埚
crumple zone 防撞压损区

cryo(e)lectronics 低温电子学
cryogenic engineering 低温工程
cryogenic pump 低温抽气泵
cryogenics 低温学
cryohydrate 〈化〉冰盐
cryometer 低温计
cryoscopic constant 冰点降低常数
cryosistor 低温晶体管
cryosorption 低温吸附，冷凝吸附
cryostat 低温恒温器，低温保持器
cryotron 冷子管，低温管
crystal 晶体，结晶
crystal class 晶类
crystal column 晶列
crystal counter 晶体计数器
crystal defect 晶体缺陷
crystal direction 晶向
crystal direction vector 晶向矢
crystal edge 晶棱，晶缘
crystal face 晶面
crystal form 晶形
crystal growth 晶体生长
crystal habit 晶体惯态
crystal lattice 晶格，晶体点阵
crystalline grain 晶粒
crystalline solid 结晶固体
crystalline state 晶态
crystalline structure 晶体结构
crystallinity 晶性，结晶度
crystallite 微晶(体)
crystallogram 晶体衍射图
crystallographic axial angle 晶轴角
crystallographic axis 晶轴
crystallography 晶体学，结晶学
crystalloid 拟晶体，凝晶质
crystal microphone 晶体传声器
crystal nucleus 晶核
crystal optics 晶体光学
crystal oscillator 晶体振荡器
crystal physics 晶体物理学
crystal pick-up 晶体拾音器
crystal plane vector 晶面矢
crystal spectrum 晶体光谱
crystal structure 晶体结构
crystal system 晶系
CT (computerized tomography) 计算机层析成像
CU (control unit) 控制部件，控制器
cubical expansion 体积膨胀
cubic centimetre (cc) 立方厘米(体积单位)
cubic close packing 立方密堆积
cubic crystal 立方晶体
cubic equation 〈数〉三次方程
cubic system 立方晶系
curie 居里(放射性强度单位)
Curie point 居里点，居里温度
Curie's law 居里定律
Curie temperature 居里点，居里温度
curl 旋度
curl electric field 有旋电场
curl field 旋度场
current 电流；流
current balance 电流天平
current density 电流密度
current element 电流元
current intensity 电流强度
curvature 〈数〉曲率
curvature of field 像场弯曲，场曲
curvature of spacetime 时空曲率
curved spacetime 弯曲时空
curve fitting 曲线拟合
curvilinear motion 曲线运动
cusp 尖拐(型突变)；〈几何〉尖点；极隙
cusp field 会切场
cut 截断
cutoff frequency 截止频率
cutoff wavelength 截止波长
CVD (chemical vapour deposition) 化学汽相沉积
cycle 周期，循环；周
cycloid 〈几何〉摆线，轮转线
cyclotron 回旋加速器
cyclotron frequency 回旋频率
cyclotron instability 回旋不稳定性

cyclotron radiation 回旋辐射
cyclotron radius 回旋半径
cyclotron resonance 回旋共振
cylindrical harmonics 〈数〉柱谐函数,圆柱调和函数,贝塞尔函数
cylindrical lens 柱面透镜
cylindrical mirror 柱面镜
cylindrical polar coordinates 柱面极坐标
cylindrical wave 柱面波

Dd

d'Alembert equation 达朗贝尔方程
d'Alembert inertial force 达朗贝尔惯性力
d'Alembert principle 达朗贝尔原理
dalton 道尔顿(原子质量单位)
Dalton's atomic theory 道尔顿原子理论
Dalton's law of partial pressure 道尔顿分压定律
damped collision 阻尼碰撞
damped harmonic motion 阻尼谐运动
damped oscillator 阻尼振子,阻尼振荡器
damped vibration 阻尼振动,衰减振动
damping 阻尼,衰减,减幅
damping coefficient 阻尼系数,衰减系数
damping force 阻尼力
dangling bond 悬键
Daniell cell 丹聂尔电池
dark band 暗带
dark field 暗(视)场
dark line 暗线
dark matter 暗物质
dark noise 暗噪声
DAT (digital audio tape) 数字录音带
data 数据
data acquisition 数据采集
database 〈计算机〉数据库
data capacity 数据容量,信息容量
data processing 数据处理
daughter nucleus 子核
Davisson-Germer experiment 戴维孙-革末实验
day 日
dB (decibel) 分贝(耳)
DC; d.c. (direct current) 直流(电)
deactivation 去激活,去活化,钝化
deadbeat 无差拍;不摆
dead time 〈化〉死时间
deathnium 致命杂质
de Broglie relation 德布罗意关系
de Broglie wave 德布罗意波
de Broglie wavelength 德布罗意波长
debye 德拜(电偶极矩单位)
Debye-Hückel equation 德拜-休克尔方程
Debye-Hückel theory 德拜-休克尔(溶液)理论
Debye model 德拜模型
Debye temperature 德拜温度
decade resistance box 十进电阻箱
decalescence 吸热变暗
decay 衰变;衰落;衰减
decay constant 衰变常量,隐灭常数,衰变恒量
decay energy 衰变能
decaying wave 衰减波
decay law 衰变定律,衰变律
decay modes 衰变方式;衰变图
decay product 衰变产物
decay rate 衰变率,衰减率
decay width 衰变宽度
deceleration 减速,减速度

decelerator 缓动装置,减速器,制动器
decibel (dB) 分贝(耳)
decimal system 十进位制
declination 偏角;磁偏角;偏差
decoloration 脱色,去色,漂白
deconfinement 退禁闭
decoupled band 退耦带
decoupling parameter 退耦参量
deep inelastic collision 深度非弹性碰撞
deep inelastic scattering 深度非弹性散射
deep level 深能级
deep level transient spectroscopy 深能级暂态谱学
deexcitation 退激发,去激发,去激励
defect 缺陷;故障
definite integral 〈数〉定积分
deflecting coil 偏转线圈
deflecting electrode 致偏电极
deflection 偏转,偏斜;偏差;挠度
deflection function 偏转函数
defocused beam 散焦光束,散焦射束
defocused image 散焦像,离焦像
defocusing 散焦,去焦
deformation 形变,畸变;失真
deformation energy 形变能
deformation parameter 形变参量,形变参数
deformed nucleus 变形核
deformed shell model 变形壳模型
degassing 除气
degaussing 消磁,去磁,退磁
degeneracy 简并性,简并度,退化度
degeneracy criterion 简并性判据
degeneracy pressure 简并压力
degeneracy semiconductor 简并半导体
degeneracy temperature 简并温度
degenerate energy level 简并能级
degenerate gas 简并气体
degenerate quantum gas 简并量子气体
degenerate semiconductor 简并半导体
degenerate states 简并态,退化态
degenerate system 简并系统,退化系
degree 度;级;幂次
degree Celsius 摄氏度(温度单位)
degree Fahrenheit 华氏度(温度单位)
degree of coherence 相干度
degree of freedom 自由度
degree of polarization 偏振度
degree of polymerization 聚合度
dehumidifier 减湿器,干燥器
dekatron 十进计数管
delayed neutron 缓发中子
delayed proton 缓发质子
delay line 延迟线
deliquescence 潮解,溶解,液化
demagnetization 退磁,去磁
demagnetization factor 退磁因数,去磁因数
demodulation 解调;检波
dendritic growth 枝晶生长,枝蔓晶体生长
densitometer 显像密度计,光密度计,密度计
density 密度
density fluctuation 密度涨落,密度波动
density functional method 密度泛函法
density matrix 密度矩阵
depletion effect 耗尽效应
depletion layer 耗尽层
depolarization 退极化;退偏振,消偏振
depolarization factor 退极化因子;消偏振因数
deposition 沉积,淀积;蒸镀
depression of freezing point 凝固点降低
depth of field 景深,视场深度

depth of focus 焦深;震深
derivative 〈化〉衍生物;〈数〉导数,微商
derived unit 导出单位
desorption 脱附,解吸(作用)
destructive interference 相消干涉,破坏性干扰
detachment process 分离过程
detailed balancing 细致平衡
detailed balancing principle 细致平衡原理
detection 检测,检波,探测
detection efficiency 检波效率,探测效率
detection sensitivity 探测灵敏度
detector 探测器;检波器
detector resolution 探测器分辨率
detonating gas 爆鸣气
deuterated compound 重氢化合物,含氘化合物
deuterium 氘,重氢
deuterium oxide 氧化氘,重水
deuteron 氘核,重氢核
deviation 偏差;〈数〉偏差数;偏转,漂移
device 器件,装置
devil's staircase 魔(鬼楼)梯
dew 露
Dewar flask 杜瓦(真空)瓶
dew point 露点
dew-point hygrometer 露点湿度计
dextrorotatory 右旋性的
diamagnetism 抗磁性,逆磁性,反磁性学
diamond 金刚石
diaphragm 光阑,光圈
diastereoisomers 非对映异构物
diathermancy 透热性
diatomic gas 双原子气体
dibaryon 双重子
dichroic mirror 分色镜,分光镜;二向色镜
dichroism 二向色性,二色性,分色性
dichromatism 二色性
dielasticity 逆弹性
dielectric 电介质,介电体
dielectric aging 介电老化
dielectric anisotropy 介电各向异性
dielectric breakdown 介电体击穿
dielectric ceramics 介电陶瓷
dielectric constant 介电常量,介电常数,电容率
dielectric constant of vacuum 真空介电常量
dielectric heating 电介质加热
dielectric loss 介电损耗,介电损失
dielectric polarization 介电极化,电介质极化
dielectric strength 介电强度,介质强度,绝缘强度
dielectric susceptibility 介电极化率
diesel engine 柴油机
difference frequency 差频
differential amplifier 差分放大器
differential calculus 〈数〉微分学
differential equation 微分方程
differential gear train 〈机械〉差动齿轮系
differential geometry 〈数〉微分几何
differential scattering cross-section 微分散射截面
differential thermometer 差示温度计
differentiation 〈数〉微分法
diffracted wave 衍射波
diffraction 衍射,绕射
diffraction angle 衍射角
diffraction fringe 衍射条纹,绕射条纹
diffraction grating 衍射光栅
diffraction optics 衍射光学
diffraction pattern 衍射图样
diffraction screen 衍射屏
diffractometer 衍射仪,衍射计
diffused light 漫射光

diffuser 漫射体；扩散器；扩压器
diffuse reflection 漫反射
diffusion 扩散；漫射
diffusion cloud chamber 扩散云室
diffusion kernel 扩散核函
diffusion length 扩散长度
diffusion limited aggregation (DLA) 扩散置限聚集
diffusion pump 扩散泵
diffusivity 扩散能力，扩散率
digit 数字，数位
digital audio tape (DAT) 数字录音带
digital computer 数字计算机
digital display 数字显示
digital frequency meter 数字频率计
digital hologram 数字全息图
digital multimeter 数字式多用表
digital recording 数字录制
digital timer 数字计时器
digital voltmeter 数字伏特计
digitron 数字指示管，数字读出辉光管
dihedron 〈数〉二面体
dilatancy 膨胀
dilatation 膨胀
dilation 膨胀
dilatometer 膨胀计
dimension 量纲；维度；尺寸
dimensional analysis 量纲分析，因次分析
dimuon 双 μ 子
diode 二极管
diode laser 二极管激光器
dioptre 屈光度，折光度
dioptrics 折射光学，屈光学
dip 倾斜度，倾角
dipole 偶极子，偶极
dipole approximation 偶极近似
dipole-dipole interaction 偶极-偶极相互作用
dipole moment 偶极矩
dipole radiation 偶极辐射
diquark 双夸克
Dirac equation 狄拉克方程
direct absorption 直接吸收
direct band gap 直接带隙
direct correlation 直接关联
direct current (DC; d.c.) 直流(电)
direct current bridge 直流电桥
direct current motor 直流马达，直流电动机
directed-beam 定向光束
direct impact 正碰
directional aerial 定向天线
directive aerial 定向天线
direct motion 〈天文〉顺行
directrix 〈几何〉准线
direct vision prism 直视棱镜
discharge 放电
discharge resistance 放电电阻
discharge tube 放电管
discontinuous function 〈数〉间断函数
discrete eigenvalue 离散本征值，分立本征值
discrete fluid model 离散流体模型
discrete picture 离散图像
discrete spectrum 离散谱，分立谱
discrimination 鉴别
discriminator 〈无线电〉鉴别器，鉴频器
disintegration 蜕变
disintegration constant 蜕变常数，蜕变恒量
diskette 〈计算机〉软盘，磁盘
dislocation 位错，断层
disorder 无序
dispersed phase 弥散相，分散相
dispersing prism 色散棱镜
dispersion 色散，弥散
dispersion curve 色散曲线
dispersion equation 色散方程
dispersion force 分散力
dispersion relation 色散关系
dispersive medium 色散介质，色散媒质

dispersive power 色散本领,色散率
displacement 位移
displacement current 位移电流
dissipation 耗散,消耗,弥散
dissipative flow 耗散流
dissipative force 耗散力
dissipative structure 耗散结构
dissipative system 耗散系统
dissociation 离解
distance 距离,程
distance of distinct vision 明视距离
distance ratio 距离比
distance-time graph 时距图,时距关系线图
distortion 畸变,失真
distributed force 分布力
distributive law 〈数〉分配律
distributor 分配器;配电盘
div (divergence) 〈数〉散度;发散
divergent series 〈数〉发散级数
divergent wave 发散波
diverging lens 凹透镜,发散透镜
diverging mirror 凹面镜,发散镜
division of amplitude 振幅分割,分振幅法
DLA (diffusion limited aggregation) 扩散置限聚集
domain 畴;〈数〉域
donor 施主
donor-acceptor pair luminescence 施主-受主对发光
donor density 施主浓度
donor ionization energy 施主电离能
doping 掺杂
Doppler broadening 多普勒(谱线)增宽
Doppler effect 多普勒效应
Doppler shift 多普勒频移
Doppler shift attenuation method 多普勒频移衰减法
Doppler velocimeter 多普勒速度计
dose 剂量
dose equivalent 剂量当量
dosimeter 放射量测定器;剂量计
dot product 点积
double β-decay 双β衰变
double-arm spectrometer 双臂谱仪
double-beam interference 双光束干涉
double-beam spectrophotometer 双束分光光度计
double concave lens 双凹透镜
double convex lens 双凸透镜
double-core shower 双芯簇射
double-differential cross-section 双微分截面
double-doped laser 双掺杂激光器
double-elliptical cavity 双椭圆(共振)腔
double-exposure holography 两次曝光全息术
double-focusing spectrometer 双聚焦谱仪
double galaxy 〈天文〉双重星系
double-gate FET (double-gate field effect transistor) 双栅场效(应)管
double-humped barrier 双峰势垒
double-image prism 双像棱镜
double refraction 双折射
double-resonance 双共振
double-slit diffraction 双缝衍射
double-slit interference 双缝干涉
doublet 〈光〉双合透镜;〈光〉双重线;〈电磁〉偶极子;双重态
d-quark d夸克,下夸克
drag coefficient 曳引系数,曳力系数
drain 疏水;排液装置;(场效应晶体管的)漏极
drift chamber 漂移室
drift mobility 漂移迁移率,迁移率
drift velocity 漂移速度
driven damped oscillator 受驱阻尼振子
driven oscillation 受驱振荡

driving force 驱动力,策动力
dry cell 干电池
dry ice 干冰
dual-cavity laser 双腔激光器
dual field 对偶场
dual-grating spectrograph 双光栅摄谱仪
duality 二元性,对偶性,二象性,二重性
dual space 对偶空间
dual tensor 对偶张量
ductile-brittle transition 延脆转变
ductile fracture 延性断裂,韧性断裂
ductile material 延性材料
ductility 延度,延性
Dulong-Petit law 杜隆-珀蒂定律
dust core (磁)粉芯,尘核
dwarf star 〈天文〉矮星
dye laser 染料激光器
dynamical matrix 动力学矩阵
dynamical pressure 动压,动压力,动压强
dynamical symmetry 动力学对称
dynamical symmetry breaking 动力学对称破缺
dynamical system 动力学系统,动力系;动态系统
dynamical variable 动力学变量,力学量
dynamic equilibrium 动力平衡,动态平衡
dynamic range 动态范围
dynamics 动力学
dynamic scattering 动态散射
dynamometer 测力计;功率计
dynamotor 直流电动发电机
dyne 达因(力的单位)
dyon 双荷子
dyotron 超高频振荡三极管,微波三极管
dystectic mixture 高熔混合物

Ee

early universe 早期宇宙
earth 地;接地
earthquake 地震
earth's atmosphere 地球大气
earthshine 〈天文〉地球反照
earth's magnetic field 地磁场
earth wire 地线,接地线
easy axis of magnetization 易磁化轴
easy direction for magnetization 易磁化方向
ebullioscopic constant 沸点升高常数
eccentricity 〈数〉离心率
ECG (electrocardiogram) 心电图
echelette grating 小阶梯光栅
echelle grating 中阶梯光栅
echelon grating 阶梯光栅
echo 回声,回音,回波
echo chamber 回声室,回响室
echolocation 回声定位法,回声测定
echo sounder 回声探测器,回音测深器
ECL (emitter-coupled logic) 发射极耦合逻辑
eclipse 〈天文〉(日、月)食
ecliptic 〈天文〉黄道
eddy current 涡(电)流
eddy current loss 涡流损耗
edge effect 边缘效应,边际效应
edge emission 边缘发射
edge enhancement 边缘增强(效应)
Edison cell 爱迪生电池
EEG (electroencephalogram) 脑电图
EELS (electron energy loss spectrometer) 电子能耗谱仪
effective aperture 有效孔径,有效口径
effective charge 有效电荷
effective conductivity 有效电导率
effective cross-section 有效截面
effective line width 有效线宽
effective mass 有效质量
effective potential 有效势
effective radius 有效半径
effective range 有效力程,有效量程,有效范围,有效距离
effective temperature 有效温度
effective value 有效值
efficiency 效率
efficiency of heat engine 热机效率
effusion 泻流,渗出;分流;隙透
EHD (electrohydrodynamics) 电流体力学
EHF (extremely high frequency) 极高频
eigenfrequency 本征频率
eigenfunction 本征函数,特征函数
eigenmode 本征模
eigen oscillation 本征振荡
eigenstate 本征态
eigenvalue 本征值,特征值
eigenvector 本征矢量,特征向量
eigenvibration 本征振动
Einstein coefficients 爱因斯坦系数
Einstein curvature tensor 爱因斯坦曲率张量

Einstein-de Haas effect 爱因斯坦-德哈斯效应
Einstein equation 爱因斯坦方程
Einstein equivalence principle 爱因斯坦等效原理
Einstein field equation 爱因斯坦场方程
Einstein frequency 爱因斯坦频率
einsteinium (Es)〈化〉锿
Einstein-Maxwell equation 爱因斯坦-麦克斯韦方程
Einstein model 爱因斯坦模型
Einstein redshift 爱因斯坦红移，引力红移
Einstein relation 爱因斯坦关系
Einstein shift 爱因斯坦位移，引力红移，谱线引力位移
Einstein temperature 爱因斯坦温度
elastance 倒电容
elastic body 弹性体
elastic collision 弹性碰撞
elastic deformation 弹性变形
elastic energy 弹性能
elastic force 弹(性)力
elasticity 弹性，弹力；弹性学
elastic modulus 弹性模数，弹性模量
elastic scattering 弹性散射
elastic scattering cross-section 弹性散射截面
elastic stress 弹性应力
elastivity 倒介电常数
elasto-optic effect 弹光效应
electret 驻极体，永电体
electric appliance 电器
electric arc 电弧
electric-arc furnace 电弧炉
electric bell 电铃
electric birefringence 电致双折射
electric bridge 电桥
electric cell 电池
electric charge 电荷
electric circuit 电路
electric conduction 电传导
electric conductivity 电导率；导电性
electric conductor (电)导体
electric current 电流
electric deflection 电偏转
electric dipole 电偶极子
electric dipole radiation 电偶极辐射
electric dipole transition 电偶极跃迁
electric displacement 电位移
electric energy 电能
electric field 电场
electric field intensity 电场强度
electric flux 电通量，电位移通量
electric flux density 电通量密度
electric force 电力
electric induction 电感应
electric insulation 电气绝缘
electricity 电；电学
electric lamp 电灯
electric leakage 漏电
electric line of force 电力线
electric motor 电动机
electric multipole 电多极子
electric multipole moment 电多极矩
electric neutrality 电中性
electric polarization 电极化强度，电极化
electric potential 电势，电位
electric potential difference 电势差，电位差
electric power 电功率，电力
electric power transmission 电力输送，输电
electric quadrupole 电四极子
electric quadrupole moment 电四极矩
electric quadrupole radiation 电四极辐射
electric quantity 电量
electric resistance 电阻
electric safety 用电安全，电气安全
electric saturation 电饱和
electric spark 电火花

electric streamline 电流线
electric susceptibility 电极化率
electric transition 电跃迁
electric wind 电风
electrification 起电;带电;电气化
electrification by friction 摩擦起电
electrified body 带电体
electrization 起电;带电;电气化
electroacoustical reciprocity theorem 电声倒易定理,电声互易原理
electroacoustic device 电声器件
electroacoustics 电声学
electroacoustic system 电声系统
electroballistics 电子弹道学
electrobiology 电生物学
electrocaloric effect 电热效应
electrocardiogram (ECG) 心电图
electrochemical cell 电化电池,化学电源
electrochemical equivalent 电化当量
electrochemical series 电化序
electrochemiluminescence 电化学发光
electrochemistry 电化学
electrochromatography 电色谱法,电色层分离法
electrocution 电击致死
electrode 电极
electrodeposit 电解淀积物
electrodialysis 电渗析
electrodynamics 电动力学
electroelastic constant 电弹常量
electroencephalogram (EEG) 脑电图
electroencephalography 脑电图学,脑电描记法
electroforming 电铸
electrohydrodynamics (EHD) 电流体力学
electrokinetics 动电学
electroluminescence 电致发光,场致发光
electrolysis 电解
electrolyte 电解质,电解(溶)液
electrolytic capacitor 电解电容
electrolytic cell 电解池,电解箱
electrolytic corrosion 电解腐蚀
electrolytic gas 电解气,爆鸣气
electrolytic rectifier 电解式整流器
electrolytic refining 电解精炼,电解精制
electrolytic separation 电解分离
electromagnet 电磁体,电磁铁
electromagnetic damping 电磁阻尼
electromagnetic decay 电磁衰变
electromagnetic energy 电磁能量
electromagnetic field 电磁场
electromagnetic field tensor 电磁场张量
electromagnetic induction 电磁感应
electromagnetic interaction 电磁相互作用
electromagnetic mass 电磁质量
electromagnetic momentum 电磁动量
electromagnetic multipole radiation 电磁多极辐射
electromagnetic pump 电磁泵
electromagnetic radiation 电磁辐射
electromagnetic radius 电磁半径
electromagnetics 电磁学
electromagnetic theory of light 光的电磁理论
electromagnetic units (e. m. u.) 电磁单位
electromagnetic wave 电磁波
electromagnetic wave spectrum 电磁波谱
electromagnetism 电磁;电磁学
electromer 电子异构体
electrometallurgy 电冶金学
electrometer 静电计
electrometer of field strength 场强计,电场强度计
electromotive force (e. m. f.) 电

动势

electromotive series 电动(次)序,电动势序列,电化序,电位序

electron 电子

electron accelerator 电子加速器

electron affinity 电子亲和性,电子亲和势

electron avalanche ionization 电子雪崩电离

electron beam 电子束

electron camera 电子摄像机

electron capture 电子俘获

electron cloud 电子云

electron diffraction 电子衍射

electron diffraction technique 电子衍射技术

electron diffractometer 电子衍射仪

electronegative 负电的,带负电的,阴电的

electronegativity 负电性,阴电性

electron energy analyzer 电子能量分析器

electron energy loss spectrometer (EELS) 电子能耗谱仪

electron energy loss spectroscopy 电子能耗谱学

electroneutrality 电中性

electron gas 电子气

electron gun 电子枪

electronic control 电子控制

electronic noise 电子噪声

electronics 电子学

electronic switch 电子开关

electronic voltmeter 电子电压表,电子伏特计

electron lens 电子透镜

electron micrograph 电子显微照片

electron microscope 电子显微镜

electron multiplier 电子倍增器

electron optics 电子光学

electron oscillation 电子振荡

electron paramagnetic resonance (EPR) 电子顺磁共振

electron phase transition 电子相变

electron-phonon interaction 电子-声子互作用

electron-positron collider 正负电子对撞机

electron-positron pair creation 正负电子对产生

electron probe microanalysis (EPM) 电子探针微区分析,电子探针微量分析

electron spectroscopy for chemical analysis (ESCA) 化学分析电子能谱法

electron spin 电子自旋

electron-spin resonance (ESR) 电子自旋共振

electron stimulated desorption 电子受激脱附

electron synchrotron 电子同步加速器

electron telescope 电子望远镜

electron tube 电子管

electronvolt (eV) 电子伏(特)

electrooptical coefficient 电光系数

electrooptical crystal 电光晶体

electrooptical deflection 电光偏转

electrooptical effect 电光效应

electrooptical modulation 电光调制

electrooptics 电光学

electroosmosis 电渗现象,电渗透

electrophile 亲电体

electrophoresis 电泳

electrophorus 起电盘

electroplating 电镀,电镀术

electropositive 正电性的,阳电性的

electroscope 验电器

electrostatic accumulation 静电积累

electrostatic agglomeration in space 空间静电凝集

electrostatic energy 静电能

electrostatic equilibrium 静电

平衡
electrostatic field 静电场
electrostatic focusing 静电聚焦
electrostatic force 静电力
electrostatic generator 静电发生器
electrostatic hazard 静电灾害
electrostatic induction 静电感应
electrostatic leakage 静电泄漏
electrostatic lens 静电透镜
electrostatic precipitator 静电除尘器
electrostatics 静电学
electrostatic screening 静电屏蔽
electrostatic shielding 静电屏蔽
electrostatic units（e.s.u.）静电系单位
electrostriction 电致伸缩
electrotechnics 电工学
electrothermics 电热学
electrotimer 电子定时器
electro-weak interaction 电弱相互作用
electro-weak unified theory 电弱统一理论
element 元素;元件
elementary beam 基本光束,元光束
elementary charge 元电荷
elementary excitation 元激发
elementary excitation spectrum 元激发谱
elementary particle 基本粒子
elementary work 元功
elements of an orbit 轨道要素,轨道根数
elevation of boiling point 沸点升高
ELF（extremely low frequency）极低频
Elinvar 镍铬恒弹性钢
ellipse〈数〉椭圆
ellipsoid〈几何〉椭球面
ellipsoidal cavity 椭球腔
ellipsoidal mirror 椭球面镜
ellipsoid of inertia 惯量椭球
ellipsometer 椭(圆)偏(振)计
ellipsometry 椭(圆)偏(振)测量术
elliptical galaxy 椭圆星系
elliptic polarization 椭圆偏振
eluate〈化〉洗出液
eluent〈化〉洗提液
elution〈化〉洗提
elutriation 淘析,淘洗
emanation 辐射,放射;射气
emergence angle 出射角
e.m.f.（electromotive force）电动势
emission 发射
emission spectrum 发射光谱,发射谱
emissive power 发射本领,发射强度,发射功率
emissivity 发射率
emittance 发射度
emitter 发射极,发射器
emitter-coupled logic（ECL）发射极耦合逻辑
emitter follower （发)射极输出器,发射跟随器
empirical equation of state 经验物态方程
empirical formula 经验公式
e.m.u.（electromagnetic units）电磁单位
emulsion （照相)乳胶,乳状液
enantiomers 对映异构体
enantiomorphism 对映形态
endoergic 吸能的
endoscope 内窥镜
endothermic reaction 吸热反应
energy 能量
energy band 能带
energy-band structure 能带结构
energy-band theory 能带论
energy conservation 能量守恒,能量不灭
energy density 能量密度
energy flux 能流,能通量
energy flux density 能流密度
energy gap 能隙

energy level 能级
energy level diagram 能级图
energy loss 能量损失
energy spectrum 能谱
energy use 能量利用,能源消费
engine 发动机,引擎
English system 英制
enrichment 浓缩
ensemble 系综
ensemble theory 系综理论
enthalpy 焓,热函
entrance pupil 入(射光)瞳
entropy 熵
entropy balance equation 熵平衡方程
entropy criterion 熵判据
entropy increase 熵增加
entropy production 熵产生
ephemeris 〈天文〉星历表,历书
ephemeris time (ET) 历书时
epicentre 震源,地震中心,震中
epicycle 〈天文〉本轮;〈数〉周转圆
epidiascope 实物幻灯机
epitaxial deposition 外延沉积
epitaxial growth 外延生长,取向附生
epitaxy 外延,取向附生
epithermal neutron 超热中子
EPM (electron probe microanalysis) 电子探针微区分析,电子探针微量分析
EPR (electron paramagnetic resonance) 电子顺磁共振
EPROM (erasable programmable read-only memory) 可擦编程只读存储器
equal inclination fringes 等倾条纹
equal inclination interference 等倾干涉
equal-spin-pairing state (ESP state) 同向自旋配对态
equal thickness fringes 等厚条纹
equal thickness interference 等厚干涉
equation of motion 运动方程
equation of state 物态方程
equation of time 时差
equator 赤道
equiconcave lens 等凹透镜
equilibrium 平衡
equilibrium condition 平衡条件
equilibrium condition of phase transition 相变平衡条件
equilibrium configuration 平衡位形
equilibrium deformation 平衡形变
equilibrium of forces 力的平衡
equilibrium position 平衡位置
equilibrium state 平衡态
equinoxes 二分点(黄道和天赤道的两个交点)
equipartition of energy 能量均分
equipartition theorem 能量均分定理
equipment 装置
equipotential line 等势线,等位线
equipotential surface 等势面
equivalence principle 等效原理
equivalent circuit 等效电路
equivalent electron 等效电子,同科电子
equivalent focal length 等效焦距
equivalent force system 等效力系
erasable programmable read-only memory (EPROM) 可擦编程只读存储器
erect image 正像
erecting prism 正像棱镜
erg 尔格(功的单位)
ergodic hypothesis 遍历假说
ergodicity 遍历性
ergonomics 人机工程学,工效学
ergosphere 〈天文〉能层
error 误差
Es (einsteinium) 〈化〉锿
Esaki diode 江崎二极管,隧道二极管
ESCA (electron spectroscopy for chemical analysis) 化学分析电

子能谱法
escape depth 逃逸深度
escapement 逃逸;擒纵机构
escape velocity 逃逸速度
ESP state (equal-spin-pairing state) 同向自旋配对态
ESR (electron-spin resonance) 电子自旋共振
estimation 估计
e.s.u. (electrostatic units) 静电系单位
ET (ephemeris time) 历书时
ether 以太;醚
ether drag 以太曳引
ether drift 以太漂移
ether wind 以太风
Euclidean dimension 欧几里得维数
eudiometer 测气管,气体燃化计
Euler equation 欧拉方程
Euler equations for hydrodynamics 欧拉流体动力学方程
Eulerian angle 欧拉角
Eulerian kinematical equations 欧拉运动学方程
eutectic mixture 共晶混合物;低共熔混合物
eutectic transformation 共晶转变
eV (electronvolt) 电子伏(特)
evaporation 蒸发;蒸镀
evaporation model 蒸发模型
evaporation spectrum 蒸发谱
even-even nucleus 偶偶核
even-odd nucleus 偶奇核
event horizon 事件视界,事件穹界
evolute 〈数〉渐屈线,法包线
evolution 演化;开方
evolution operator 演化算符
evolution rule 演化规则
exactly solved model 严格解模型
EXAFS (extended X-ray absorption fine structure) 扩展X射线吸收精细结构
excess electron 超额电子,过剩电子
excess entropy 逾熵,过剩熵
excess entropy balance equation 逾熵平衡方程
excess entropy production 逾熵产生
excess flux 逾流
excess force 逾力
exchange energy 交换能
exchange force 交换力
exchange interaction 交换作用
exchange symmetry 交换对称性
excimer 受激二聚物,受激(准分)子
excimer laser 激基分子激光器,准分子激光器
exciplex 复合受激态
exciplex laser 激基分子激光器,准分子激光器
excitation 激发,励磁
excitation curve 激发曲线,励磁曲线
excitation energy 激发能
excitation function 激发函数
excited state 激发态,受激态
exciton 激子,激发子
exciton model 激子模型
exciton spectrum 激子光谱
exclusion principle 排他原理,不相容原理
exclusive process 遍举过程
exhaust 排气;排气装置
exitance 出射度
exit window 出射窗
exoergic 放能的
exosphere 外大气层,外逸层
exothermicity 放热性
exothermic reaction 放能反应
exotic atom 奇特原子
exotic molecule 奇特分子
exotic state 奇特态
expansion 膨胀;扩展
expansion of the universe 宇宙膨胀
expansivity 膨胀率,扩展性,延伸性

expectation value 期望值,期待值
experiment 实验
experimental physics 实验物理学
experimental Q-value 实验 Q 值
explosive galaxy 爆发星系
exponent 指数
exponential growth 指数增长
exposure 曝光;曝光量,照射量
exposure meter 曝光表
extended state 扩展态
extended X-ray absorption fine structure (EXAFS) 扩展 X 射线吸收精细结构
extensibility 延伸率;延伸性
extensive quantity 广延量
extensive variable 广延变数
extensometer 伸长计,引伸计,张力计
external force 外力
external galaxy 〈天文〉河外星系
external reflection 外反射
extinction 消光;熄灭
extinction coefficient 消光系数
extinction correction 消光校正
extinction index 消光率
extinction ratio 消光比
extinction theorem 消光定理
extraction 萃取,提取;〈数〉开方
extraction potential 萃取势
extraordinary light 非常光
extraordinary ray 非常射线,非常光线
extraordinary refractive index 非常折射率
extraordinary wave 非常波
extrapolation 〈数〉外推
extremely high frequency (EHF) 极高频
extremely low frequency (ELF) 极低频
extreme path 极端光程,极远光程
extreme ultraviolet 极端紫外,极远紫外
extrinsic absorption 外赋吸收,非本征吸收
extrinsic semiconductor 外赋半导体,非本征半导体
eye 眼睛
eye lens 接目镜
eyepiece 目镜
eyepiece spectroscope 目镜分光镜
eyepoint 出射点
eye relief 眼点距,眼睛间隙

Ff

F (farad) 法拉(电容单位)
Fabry-Pérot etalon 法布里-珀罗标准具,法布里-珀罗标准器
Fabry-Pérot filter 法布里-珀罗滤波器
Fabry-Pérot interferometer 法布里-珀罗干涉仪
face-centred cubic (f.c.c.) 面心立方体
face-centred lattice 面心格
facet growth 小面生长
faceting 小面化
factor 因数,系数,因子,因素
factorial 〈数〉阶乘积,阶乘
factorization 因子化,因子分解,因式分解
factorization condition 因子分解条件
factor of quality 品质因数
Fahrenheit 华氏温度计
Fahrenheit scale 华氏温标
Fahrenheit temperature scale 华氏温标
Fahrenheit thermometric scale 华氏温标
fallout 沉降物;散落物
false neutron 非聚变中子,非热核中子
false vacuum 伪真空
family 〈化〉族
farad (F) 法拉(电容单位)
Faraday balance 法拉第磁秤,法拉第扭秤
Faraday cage 法拉第笼,法拉第屏蔽罩
Faraday constant 法拉第常数,法拉第恒量
Faraday cylinder 法拉第圆筒
Faraday effect 法拉第(涡流)效应
Faraday rotation 法拉第旋转
Faraday's laws 法拉第定律
Faraday's laws of electromagnetic induction 法拉第电磁感应定律
far field 远场
far field condition 远场条件
far field pattern 远场图样,远场图形
far infrared 远红外,远红外线
far sight 远视
fast breeder 增殖反应堆
fast breeder reactor 快中子增殖反应堆
fast Fourier transform (FFT) 快速傅里叶变换
fast neutron 快中子
fast process 快过程
fast pulley 固定轮
fast reactor 快堆
fast variable 快变量
fathometer 回音测深仪,水深计
fatigue 疲劳,疲乏
fatigue limit 疲劳限度,疲乏极限
f.c.c. (face-centred cubic) 面心立方体
FED (field-effect device) 场效应器件
feedback 反馈
FEM (field-emission microscope) 场致发射显微镜
Fermat('s) principle 费马原理
fermi 费米(长度单位)
Fermi constant 费米常数
Fermi-Dirac integral 费米-狄拉

克积分
Fermi-Dirac statistics 费米-狄拉克统计法
Fermi function 费米函数
Fermi interaction 费米相互作用
Fermi level 费米能级
Fermi momentum 费米动量
fermion 费米子
fermionic string 费米子弦
Fermi sphere 费米球
Fermi surface 费米面
fermium (Fm)〈化〉镄
ferrielectrics 亚铁电体
ferrimagnetism 亚铁磁性,铁氧体磁性
ferrite 铁素体;铁氧体
ferroalloy 铁合金
ferroelastic domain 铁弹畴
ferroelasticity 铁弹性
ferroelastics 铁弹体
ferroelectric 铁电体;铁电的
ferroelectric domain 铁电畴
ferroelectric domain wall 铁电畴壁
ferroelectric ferromagnet 铁电铁磁体
ferroelectricity 铁电性,铁电现象
ferroelectric material 铁电材料
ferroelectric sensor 铁电传感器
ferromagnet 铁磁体
ferromagnetic Curie point 铁磁居里点
ferromagnetic glass 铁磁玻璃
ferromagnetic material 铁磁材料
ferromagnetic phase transition 铁磁相变
ferromagnetic resonance (FMR) 铁磁共振
ferromagnetism 铁磁性
ferromagnon 铁磁波子,铁磁振子
fertile material 增殖材料,增殖性物质
FET (field-effect transistor) 场效(应)晶体管
few-body problem 少体问题
Feynman diagram 费恩曼图
FFT (fast Fourier transform) 快速傅里叶变换
Fibonacci numbers 斐波那契数
fibre 纤维;纤维质料
fibre bundle 纤维束,纤维丛
fibre light guide 纤维光导
fibre optics 纤维光学
fibrescope 光纤(观察)镜,纤维镜
Fick law 菲克定律
fictitious charge 虚拟电荷,虚设电荷
fidelity 保真性,保真度
field 场
field charging 场致荷电
field coil 磁场线圈
field corrector 像场校正器,像场校正镜
field effect 场效应
field-effect device (FED) 场效应器件
field-effect transistor (FET) 场效应晶体管
field emission 场致发射
field-emission microscope (FEM) 场致发射显微镜
field-free space 无场空间
field-ionization microscope (FIM) 场离子显微镜,静电离子显微镜
field-ion microscope (FIM) 场离子显微镜,静电离子显微镜
field lens (向)场镜
field magnet 场磁铁
field of view 视场
field source 场源
field stop 场阑,视场光阑
field theory 场论
filament 灯丝;丝极
filament lamp 白炽灯
file[1] 文件
file[2] 锉刀
filled band 满带
filled site 实座
film badge 胶片剂量计,胶片式射线计量器
film boiling 膜沸腾,薄膜沸腾
film interference 薄膜干涉

filter 滤波器,滤光片,过滤器;过滤
filter pump (过)滤泵
filtrate 滤出液
filtration 滤波;过滤
FIM (field-ionization microscope; field-ion microscope) 场离子显微镜,静电离子显微镜
final state 末态,终态
final velocity 末速(度)
finder 取景器;瞄准器;探测仪;寻星镜
fine adjustment 细调,微调,精密校正
fine-grained density 细粒密度
fine structure 精细结构
finite rotation 有限转动
finite series 有限级数,有穷数列
first cosmic velocity 第一宇宙速度
first harmonic 基谐波
first integral 第一积分,初积分
first law of thermodynamics 热力学第一定律
first-order phase transition 一级相变
first overtone 第一泛音
first principle 第一性原理
first sound 第一声
fissile material 可裂变物质,可分裂物质
fissile nucleus 可裂变核
fissility 可裂变性
fission 裂变,分裂
fissionable nucleus 可裂变核
fission barrier 裂变势垒,裂变位垒
fission bomb 裂变弹
fission cross-section 裂变截面
fission energy 裂变能
fission fragment 裂变碎片
fission neutron 裂变中子
fission parameter 裂变参量
fission product 裂变产物,核裂产物
fission reactor 裂变堆
fission threshold 裂变阈
fission-track dating 裂变径迹定年法
fission yield 裂变产额
Fitzgerald contraction 菲茨杰拉德收缩
fixed-axis rotation 定轴转动
fixed index 固定指标
fixed point (温标)定点,不动点,固定点
fixed pulley 定滑轮
fixed star 〈天文〉恒星
fixed target 固定靶
fixer 固定器
fixing 固定
Fizeau experiment 菲佐实验
Fizeau fringe 菲佐(干涉)条纹
Fizeau's method 菲佐法
flame 火焰
flame fusion method 焰熔法
flame photometer 火焰光度计
flash photolysis 闪光光解
flash point 闪点
flat interferometer 平面干涉仪
flatness of field 像场平度,扫描场平度
flat spacetime 平直时空
flavour 味(量子数)
flavour space 味空间
flavour symmetry 味对称性
flexoelectric effect 弯电效应
flip-chip 倒装片法
flip-flop 双稳态多谐振荡器;触发电路,触发器
floatage 浮力
floating 漂浮,浮动
floating ring 浮环
floating zone melting 悬浮区熔法
floppy disk 〈计算机〉软盘
flotation balance 浮力秤
flowmeter 流量计
fluctuating force 涨落力
fluctuation 涨落,起伏,波动
fluctuation-dissipation theorem 涨落耗散定理
fluctuation field theory 涨落场论

fluctuation term 涨落项
fluid 流体,流质;射流
fluid dynamics 流体动力学
fluidics 流控学,流控技术,射流技术
fluidity 流体性,液性,流度
fluidization 液化,流化
fluid mechanics 流体力学
fluorescence 荧光
fluorescence analysis 荧光分析
fluorescent lamp 荧光灯,日光灯
fluorescent light 荧光;荧光灯
fluoroscope 荧光镜
fluoroscopy 荧光学;荧光检测
flux 通量;能流;焊剂;熔剂
flux creep 磁通蠕变
flux density 通量密度,流量密度
flux-density vector 流密度矢量
flux flow 磁通流动
flux flow resistance 磁通流阻
flux-gate magnetometer 磁通门磁强计
fluxion 磁通量子
flux jumping 磁通跳变
fluxmeter 磁通计;辐射通量计
fluxoid 类磁通,循环量子
fly's-eye lens 蝇眼透镜,复眼透镜
FM (frequency modulation) 频率调制,调频
Fm (fermium) 〈化〉镄
FMR (ferromagnetic resonance) 铁磁共振
f-number f数,光圈号数
focal length 焦距
focal plane 焦面,焦平面
focal point 焦点
focal power (光)焦度,倒焦距
focal ratio f数,光圈数,焦比
focal tolerance 焦点容限
focometer 焦距计
focus 焦点;聚焦,调焦
focusing electrode 聚焦电极
focus in image space 像方焦点,第二焦点
focus in object space 物方焦点,第一焦点
Fokker-Planck equation 福克尔-普朗克方程
folded cavity 折叠腔
folded laser 折叠(腔)激光器
folded potential 折叠势
folded spectrum 折叠谱
foot 英尺(长度单位);底座,支点
forbidden band 禁带
forbidden emission 禁戒发射
forbidden transition 禁戒跃迁
force 力
forced convection 强制对流,强迫对流
forced light scattering 受迫光散射
forced mode-locking 强制锁模
forced vibration 受迫振动,强迫振动
force field 力场
force-free field 无力场
force multiplier 力倍增器
force ratio 力比
force screw 力螺旋
fore-prism 前置棱镜
formation factor 形成因子
formation time 形成时间
form factor 形状因子;波形因数
formula 公式
Fortin barometer 福丁气压计
forward scattering 前向散射
Foucault pendulum 傅科摆
four-acceleration 四维加速度
four-axis coordinate system 四轴坐标系
four-dimensional form of Maxwell equations 麦克斯韦方程的四维形式
four-dimensional spacetime 四维时空
four-force 四维力
Fourier analysis 傅里叶分析,(变换)谐波(量)分析
Fourier law 傅里叶定律
Fourier optics 傅里叶光学
Fourier series 傅里叶级数

Fourier synthesis 傅里叶综合(法)
Fourier transformation 傅里叶变换
four-level system 四能级系统
four-momentum 四维动量
four-stroke cycle 四冲程循环
four-stroke-cycle engine 四冲程引擎
four-tensor 四维张量
fourth dimension 第四维
four-vector 四维矢量,四元矢量
four-velocity 四维速度,四元速度
fractal 分形,分形体
fractal dimension 分形维数,分维
fractal measure 分形测度
fracture strength 断裂强度
fragmentation reaction 碎裂反应,裂片反应
frame of axes 坐标系
frame of reference 参照系,参考系
Fraunhofer diffraction 夫琅禾费衍射
Fraunhofer hologram 夫琅禾费全息图
Fraunhofer line 夫琅禾费谱线
free charge 自由电荷
free electron 自由电子
free-electron laser 自由电子激光器
free-electron model 自由电子模型
free energy 自由能
free energy criterion 自由能判据
free enthalpy 自由焓
free enthalpy criterion 自由焓判据
free expansion 自由膨胀
free fall 自由降落,自由下落
free-ion model 自由离子模型
free magnetic charge 自由磁荷
free path 自由程
free space 自由空间
free vector 自由矢量
freeze drying 冷冻干燥
freezing 凝固
freezing mixture 〈化〉冷冻混合物,冷凝剂
freezing point 凝固点,凝点,凝固温度
Frenkel defect 弗仑克尔缺陷
frequency 频率,周率
frequency domain 频域
frequency doubling 倍频
frequency down-conversion 降频转换
frequency-locking 锁频
frequency matching 频率匹配
frequency meter 频率计
frequency mixing 混频
frequency modulation (FM) 频率调制,调频
frequency of operation 工作频率
frequency plane 频谱面
frequency pulling 频率牵引
frequency pushing 频率推斥
frequency shift 频移
frequency spectrum 频谱
frequency stabilization 稳频
frequency stabilized laser 稳频激光器
frequency upconversion 升频转换
fresnel 菲(涅耳)(频率单位)
Fresnel bimirror 菲涅耳双镜
Fresnel biprism 菲涅耳双棱镜
Fresnel diffraction 菲涅耳衍射
Fresnel formula 菲涅耳公式
Fresnel lens 菲涅耳透镜
friction 摩擦,摩擦力
friction force 摩擦力
fringe 条纹;边缘
fringe contrast 条纹衬比度,(干涉)条纹反差
fringe visibility 条纹可见度
frost 霜
froth flotation 泡沫浮选
frustum 〈几何〉平截头台
fuel 燃料
fuel cell 燃料电池
fuel element 燃料元件
fugacity 逸度,逸性

fulcrum 支点，支轴
full aperture 全孔径，全口径
full coherence 完全相干性
full-colour image 全色图像
full energy peak 全能峰
full view 全景，全视图
full view hologram 全景全息图
full-wave plate 全波片
full-wave rectifier 全波整流器
function 〈数〉函数
fundamental 基本原理
fundamental constant 基本常量，基本常数
fundamental frequency 基频
fundamental mode 基模，基谐模
fundamental series 基线系
fundamental unit 基本单位
fundamental vibration 基本振动
fundamental wave 基波
fuse 熔断器，熔丝，保险丝
fuse box 熔断器盒，保险丝盒
fusible alloy 易熔合金
fusion 熔化；聚变；熔接
fusion bomb 热核弹（尤指氢弹）
fusion point 熔点
fusion reaction 熔合反应，聚变反应
fusion reactor 聚变反应堆
future 未来

Gg

g (gram) 克(质量单位)
GaAs (gallium arsenide)〈化〉砷化镓
GaAs laser 砷化镓激光器
gain 增益
gain coefficient 增益系数
gain saturation 增益饱和
gal (gallon) 加仑(容量单位)
galactic cluster 〈天文〉疏散星团,银河星团
galaxoid 〈天文〉星系体
Galaxy 〈天文〉银河系
galaxy 〈天文〉星系
Galilean invariance 伽利略不变性
Galilean principle of relativity 伽利略相对性原理
Galilean telescope 伽利略望远镜
Galilean transformations 伽利略变换
gallium arsenide (GaAs)〈化〉砷化镓
gallon (gal) 加仑(容量单位)
galvanic cell 原电池,伽伐尼电池
galvanometer 电流计,检流计,灵敏电流计
gamma decay γ衰变
gamma radiation γ辐射
gamma radioactivity γ放射性
gamma ray γ射线
gamma-ray astronomy γ射线天文学
gap level 隙能级
gas 气体
gas chromatography 〈化〉气相色谱分析;气相色谱学
gas constant 气体常数
gas-cooled reactor 气冷反应堆
gas dynamics 气体动力学
gas kinetics 气体动理学,气体分子运动论
gas laser 气体激光器
gas laws 气体定律
gas-liquid continuity 气液连续性
gasoline engine 汽油机
gas sensor 气敏器件,气体敏感器
gas thermometer 气体温度计
gas turbine 燃气轮机
gate 〈计算机〉门,门电路
gauge boson 规范玻色子
gauge field 规范场
gauge field theory 规范场论
gauge invariance 规范不变性
gauge particle 规范粒子
gauge potential 规范势
gauge transformation 规范变换
gauss 高斯(磁感应强度单位)
Gaussian beam 高斯光束
Gaussian curvature 高斯曲率
Gaussian distribution 高斯分布,正态分布
Gaussian optics 高斯光学
Gaussian units 高斯单位
Gauss' law 高斯定律
gaussmeter 高斯计,磁强计
Gauss model 高斯模型
Gauss(') theorem 高斯定理
Gay-Lussac's law 盖-吕萨克定律
Gb (gilbert) 吉伯(磁通势单位)
Ge (germanium)〈化〉锗
gear 齿轮,传动装置
gedanken experiment 理想实验

Gegenschein 对日照
Geiger counter 盖革计数器
Geiger-Müller counter 盖革-米勒计数器
Geissler tube 盖斯勒管
gel 凝胶
gelation 凝胶化
gelation point 凝胶化点
gel phase 凝胶相
generalized force 广义力
generalized Laguerre polynomial 广义拉盖尔多项式
generalized Langevin equation 广义朗之万方程
generalized momentum 广义动量
generalized Ohm law 广义欧姆定律
generalized velocity 广义速度
general physics 普通物理学
general theory of relativity 广义相对论
generation time 产生时间
generator 发电机;发生器
generatrix 母线
geocentricism 地球中心说
geocentric system 地心体系
geocentric theory 地球中心说,地心说
geodesic 大地线,测地线
geodesic line 大地线,测地线
geodesy 测地学
geomagnetic field 地磁场
geomagnetism 地磁;地磁学
geometrical aberration 几何像差
geometrical optics 几何光学
geometric average 几何平均数
geometric series 几何级数,等比级数
geometrized units 几何化单位
geophysics 地球物理学
geostationary orbit 对地静止轨道,同步轨道
geosynchronous orbit 同步轨道
geothermal energy 地热能
germanium (Ge) 〈化〉锗
German silver 德银,铜镍锌合金
getter 吸气剂,收气剂;吸杂剂
GeV (gigaelectronvolt) 千兆电子伏特,十亿电子伏特
giant star 〈天文〉巨星
gibbous Moon 〈天文〉凸月
Gibbs free energy 吉布斯自由能
Gibbs function 吉布斯函数
Gibbs paradox 吉布斯佯谬
gigaelectronvolt (GeV) 千兆电子伏特,十亿电子伏特
gilbert (Gb) 吉伯(磁通势单位)
gimbal 万向节
Giorgi system 乔吉制
glancing angle 掠射角
glancing collision 小角度碰撞
glancing incidence 掠入射,小角度入射
Glashow-Weinberg-Salam model (GWS model) GWS模型,格拉肖-温伯格-萨拉姆模型
glass 玻璃
glass fibre 玻璃纤维
glass transition 玻璃化转变,琉态转变
glass transition temperature 玻璃化转变温度
glide plane 滑移面
glide reflection 滑移反射
globe lens 球透镜
globular cluster 〈天文〉球状星团
glove box 手套箱
glow 辉光
glow curve 辉光曲线
glow discharge 辉光放电
glueball 胶球
gluino 胶微子
gluon 胶子
Golay cell 戈莱盒
Goldstone particle 戈德斯通粒子
Goldstone('s) theorem 戈德斯通定理
good conductor 良导体
good quantum number 好量子数
governor 调速器
G-parity G宇称

graded heterojunction 缓变异质结
graded junction 缓变结
gradient 梯度,倾斜度
gradient B drift 磁场梯度漂移
gradient expansion 梯度展开
gradient operator 梯度算子
Graham('s) law 格拉姆定律
grain boundary 晶界
gram (g) 克(质量单位)
gram calorie 克卡(热量单位)
grand canonical distribution 巨正则分布
grand canonical ensemble 巨正则系综
grand partition function 巨配分函数
grand potential 巨(热力学)势
grand unification 大统一
grand unified theory (GUT) 大统一理论
graph 图,曲线图
graphical expansion 图解展开
graphical method 图解法
graphite moderator 石墨减速剂
graticule 分划板;槽纹光栅;十字线;分度镜
grating 光栅
grating spectrograph 光栅摄谱仪
gravitation 引力
gravitational collapse 引力坍缩
gravitational condensation 引力凝聚
gravitational constant 引力常量
gravitational drift 重力漂移
gravitational field 引力场,重力场
gravitational field strength 引力场强度
gravitational force (万有)引力,重力
gravitational interaction 引力相互作用
gravitational lens 引力透镜
gravitational mass 引力质量
gravitational potential 引力势,重力势
gravitational potential energy 引力位能
gravitational radiation 引力辐射
gravitational radius 引力半径
gravitational redshift 引力红移,爱因斯坦红移
gravitational shift 引力移位
gravitational wave 引力波
gravitation time dilation 引力时间延缓
gravitino 引力微子
graviton 引力子
gravity 重力,地心引力
gravity field 重力场
gravity-free environment 无重力环境
gray (Gy) 戈(瑞)(吸收剂量单位)
grazing angular momentum 擦边角动量
grazing collision 擦边碰撞
great circle 大圆
Green function 格林函数
greenhouse effect 温室效应
Gregorian telescope 格雷果里望远镜
grey group 灰(色)群
grey level 灰阶
grid 栅极;格栅;电池铅板
ground 接地
ground state 基态
ground wave 地面电波
group 〈化〉族;〈数〉群
group velocity 群速,群速度
growth of single crystal 单晶生长
g-state g 态
guided wave 受导波
guiding centre 导心
guiding laser beam 制导激光束
gun metal 炮铜,铜锡合金
GUT (grand unified theory) 大统一理论

GWS model (Glashow-Weinberg-Salam model) GWS模型,格拉肖-温伯格-萨拉姆模型
Gy (gray) 戈(瑞)(吸收剂量单位)
gyrocompass 旋转罗盘,回转罗盘
gyroelectric effect 旋电效应
gyromagnetic device 旋磁器件
gyromagnetic effect 旋磁效应
gyromagnetic ratio 旋磁比
gyroradius 回旋半径
gyroscope 陀螺仪,回旋装置,回转仪

Hh

H[1] (henry) 亨利(电感单位)
H[2] (hydrogen) 〈化〉氢
hadrodynamics 强子动力学
hadron 强子
hadron calorimeter 强子量能器
hadronic decay 强子型衰变
hadronization 强子化
hair hygrometer 毛发湿度计
half cell 半电池
half-life 半衰期
half-peak width 半峰全宽
half-period zone 半周期带
half-shade device 半影器件,半影装置
half-silvered mirror 半镀银镜
half-symmetric resonator 半对称共振腔
half-time 半衰期
half-value thickness 半值厚度
half-wave aerial 半波天线
half-wave loss 半波损失
half-wave plate 半波片
half-wave rectifier 半波整流器
half-width 半宽度
half width at half maximum 半峰半宽
Hall effect 霍耳效应,内液增阻效应
Halley's comet 〈天文〉哈雷彗星
halo 〈天文〉晕;〈电子〉光晕
halogen 〈化〉卤素
Hamiltonian 哈密顿量,哈密顿函数,哈密顿算符
Hamiltonian characteristic function 哈密顿特征函数
Hamiltonian function 哈密顿函数
Hamiltonian operator 哈密顿算符
Hamilton principle 哈密顿原理
hard axis of magnetization 难磁化轴
hard direction for magnetization 难磁化方向
hard excitation 硬激发
hard magnetic bubble 硬磁泡
hard magnetic material 硬磁材料
hard mode 硬模
hardness 硬度
hard process 硬过程
hard proton 硬光子
hard radiation 超短波辐射,硬性辐射
hardware 〈计算机〉硬件
harmonic 谐波,谐音
harmonic oscillator 谐振子,谐波振荡器
harmonic progression 调和级数
harmonic series 调和级数
harmonic sound 谐音
Harvard classification 哈佛(恒星光谱)分类法
head-on collision 对头碰撞,正碰,对心碰撞
health physics 保健物理学
heat 热,热量
heat balance 热平衡
heat capacity 热容(量)
heat conduction 热传导
heat conductivity 热导率,热导性
heat death 热寂
heat energy 热能
heat engine 热机

heat exchanger 热交换器
heat insulation 热绝缘,绝热
heat of adsorption 吸附热
heat of atomization 原子化热
heat of combustion 燃烧热
heat of dissociation 离解热
heat of evaporation 蒸发热
heat of formation 形成热
heat of fusion 熔化热,熔解热
heat of neutralization 中和热
heat of reaction 反应热
heat of solution 溶解热
heat of sublimation 升华热
heat of vaporization 汽化热
heat pump 热泵
heat radiation 热辐射
heat reservoir 热库,热源
heat shield 热屏,挡热板
heat sink 热汇,冷源
heat source 热源
heat transfer 传热
Heaviside layer 亥维赛层,E 层
Heaviside-Lorentz units 亥维赛-洛伦兹单位
heavy-damped wave 高阻尼波
heavy fermion 重费米子
heavy hydrogen 氘,重氢
heavy ion 重离子
heavy-ion collision 重离子碰撞
heavy-ion radioactivity 重离子放射性
heavy-ion reaction 重离子反应
heavy lepton 重轻子
heavy nucleus 重核
heavy photon 重光子
heavy quark 重夸克
heavy water 氧化氘,重水
hectare 公顷(面积单位)
HEED (high-energy electron diffraction) 高能电子衍射
height above sea level 海拔高度
Heisenberg uncertainty principle 海森伯测不准原理
helicity 螺旋性,螺旋度
helimagnetism 螺旋磁性
heliocentric system 日心体系
heliocentric theory 地动说,日心说
heliostat 定日镜
helium-cadmium laser 氦镉激光器
helium-neon laser (He-Ne laser) 氦氖激光器
helium refrigerator 氦冷冻机,氦制冷机
Helmholtz coils 亥姆霍兹线圈
Helmholtz equation 亥姆霍兹方程
Helmholtz free energy 亥姆霍兹自由能
Helmholtz-Lagrange theorem 亥姆霍兹-拉格朗日定理
He-Ne laser (helium-neon laser) 氦氖激光器
henry (H) 亨利(电感单位)
Henry's law 亨利定律
Hermite-Gauss beam 厄米-高斯光束
Hermite-Gauss mode 厄米-高斯模,厄米-高斯波型
Hermitian conjugate 厄米共轭
Hermitian operator 厄米算符
hertz (Hz) 赫兹(频率单位)
Hertzian oscillator 赫兹振子
Hertzsprung-Russell diagram (H-R diagram) 赫罗图
Hess's law 赫斯定律
heterocharge 异极电荷,混杂电荷
heterochromatic light 杂色光
heterodyne 外差的,外差法的
heteroepitaxy 异质外延
heterogeneous light 杂色光
heterojunction 异质结
heterojunction laser 异质结激光器
heterolaser 异质结激光器
heterostructure 异质结构,异质结结构
Heusler alloys 郝斯勒合金
hexagonal close packing 六角密堆积

hexagonal system 六角晶系
HF (high frequency) 高频
hidden symmetry 隐对称性
hidden variable 隐变量
hi-fi (high fidelity) 高保真度
Higgs field 希格斯场
Higgs particle 希格斯粒子
high-aperture lens 大孔径透镜
high-energy astrophysics 高能天体物理学
high-energy electron diffraction (HEED) 高能电子衍射
high-energy nuclear physics 高能核物理学
high-energy nuclear reaction 高能核反应
high-energy physics 高能物理学
high fidelity (hi-fi) 高保真度
high frequency (HF) 高频
high-frequency discharge 高频放电
high-frequency plugging 高频堵漏
high-reflecting film 高反射膜
high-resolution spectroscopy 高分辨光谱学
high-speed camera 高速照相机,高速摄影机
high-speed steel 高速钢
high-temperature plasma 高温等离子体
high tension (HT) 高电压
high vacuum 高真空
high vacuum insulation 高真空隔热
Hilbert space 希尔伯特空间
Hildebrand rule 希尔德布兰德规则
hindrance factor 阻碍因子
hoarfrost 霜,白霜
hole 空穴,孔
hole state 空穴态
hologram 全息图
holography 全息摄影术,全息术
holophote 全聚反光装置
homocentric beam 同心光束,共心光束
homocharge 同极电荷
homoepitaxy 同质外延,均相外延
homojunction 同质结
homojunction laser 同质结激光器
Hooke's law 胡克定律
horizontal projection 水平投影
horsepower (hp) 马力(功率单位)
hot electron 过热电子
hot luminescence 过热发光
hot plasma 热等离子体
hot-wire instrument 热线式电气测试仪表
hour 小时(时间单位)
howl 啸叫
hp (horsepower) 马力(功率单位)
H-R diagram (Hertzsprung-Russell diagram) 赫罗图
HT (high tension) 高电压
Hubble constant 哈勃常数
Hubble law 哈勃定律
Hubble space telescope 哈勃天文望远镜
hue 色调,色彩
hum 哼鸣,哼扰
humidification 增湿
humidifier 增湿器,湿润器
humidistat 湿度调节器
humidity 湿度
Humphrey series 汉弗莱级数
Hund's rules 洪德定则
hunting 搜索;寄生振荡,不规则振荡
Huygens(') construction 惠更斯作图法
Huygens(') principle 惠更斯原理
hybrid 混杂子
hydrated crystal 含水晶体
hydraulic press 水压机
hydraulic pressure 水压
hydraulics 水力学
hydrodynamics 流体动力学
hydroelectricity 水电,水力电气
hydrogen (H) 〈化〉氢
hydrogen arc lamp 氢弧灯

hydrogen bomb 氢弹
hydrogen bond 氢键
hydrogen electrode 氢电极
hydrogen half cell 氢半电池
hydrogen ion 氢离子
hydrogen-like atom 类氢原子
hydrogen spectrum 氢光谱
hydromechanics 流体力学
hydrometeor 水汽凝结体;水汽现象
hydrometer 液体比重计,比重计,浮秤
hydrostatic pressure 流体静压强,流体静压力
hydrostatics 流体静力学
hygrometer 湿度计
hyperbola 双曲线
hyperbolic function 双曲函数
hyperchaos 超混沌
hypercharge 超荷
hyperchromatic lens 多色差透镜
hyperfine splitting 超精细分裂
hyperfine structure 超精细结构
hyper(metr)opia 远视
hyperon 重核子,超子
hypersonics 特超声学
hypertext 〈计算机〉超文本
hypertonic solution 高渗(压)溶液
hypothesis 假设,假说
hypotonic solution 低渗溶液
hypsometer 测高计
hysteresis 滞后(效应),滞后作用
hysteresis curve 滞后曲线,磁滞曲线
hysteresis loop 滞后回线,磁滞回线
Hz (hertz) 赫兹(频率单位)

Ii

IBA (ion-beam analysis) 离子束分析
IC (integrated circuit) 集成电路
ice 冰
ice condition 冰条件
ice crystal 冰式晶体,冰晶
ice entropy 冰熵
ice model 冰模型
ice point 冰点
iconoscope 光电显像管,光电摄像管
icosahedral group 二十面体群
icosahedron 二十面体
ideal constraint 理想约束
ideal crystal 理想晶体
ideal fluid 理想流体
ideal gas 理想气体
ideal gas thermometric scale 理想气体温标
ideal solution 理想溶液,完美溶液
identical particles 全同粒子
identity 恒等式,恒等,同一性
IF (intermediate frequency) 中频
IGFET (insulated-gate field effect transistor) 绝缘栅场效(应)管
ignition 点火,引燃
ignition energy 点火能
ignition temperature 点火温度
illuminance 照度
illumination 照度,照明
illuminometer 照度计
image 像;图像;映像
image blurring 图像模糊
image construction 求像(作图)法
image conversion 图像变换
image converter 像转换器
image deblurring 图像去模糊
image definition 图像清晰度,图像分解力,(底片的)反差度
image distance 像距,像主距
image element 像元,像素
image encoding 图像编码
image enhancement 图像增强
image field 像场
image intensifier 像增强器,像管
image plane 像平面
image processing 图像处理
image quality 像质
image restoration 图像复原
image space 像(方)空间
image storage 图像存储
image synthesis 图像合成,图像综合
imaginary number 〈数〉虚数
imaging 成像
immersion objective 浸没物镜
impact broadening 碰撞(谱线)增宽,碰撞展宽
impact ionization 碰撞电离
impact parameter 碰撞参量,碰撞参数
impact printer 击打式打印机
impedance 阻抗
impedance matching 阻抗匹配
imperfect elastic collision 非完全弹性碰撞
imperfect gas 非理想气体
Imperial units 英制单位
implantation 植入
implosion 爆聚,向心爆炸;暴缩
implosion heating 爆聚加热
improper ferroelectrics 非本征铁

电体
impulse 冲量;脉冲
impulse response 脉冲响应
impulse strength 脉冲强度
impulse voltage 脉冲电压
impurity 杂质
impurity level 杂质能级
impurity resistivity 杂质电阻率
incandescence 白热,炽热
incandescent lamp 白炽灯
inch 英寸(长度单位)
incidence 入射
incidence angle 入射角,迎角
incident particle 入射粒子
incident ray 入射线
incident wave 入射波
inclination 倾度,倾角,磁倾角
inclination factor 倾斜因子,倾斜因数
inclined plane 斜面
inclinometer 磁倾计,倾斜计
incoherence 非相干性
incoherent illumination 非相干照明
incoherent imaging 非相干成像
incoherent light 非相干光
incoherent scattering 非相干散射,杂乱散射
incoming beam 入射束
incoming channel 入射道
incoming solar radiation 入日射,太阳入射辐射
incommensurate phase 无公度相
incompatibility 不相容性
incomplete fusion 非完全熔合
incompressibility 不可压缩性
indefinite integral 不定积分
independence 无关性
independent amplitude 独立振幅
independent-particle model 独立粒子模型
indeterminacy 不确定性
index theorem 指数定理
indices of crystal direction 晶向指数
indices of crystal plane 晶面指数
indifferent equilibrium 随遇平衡
indirect measurement 间接测量
indium antimonide 〈化〉锑化铟
induced charge 感生电荷,感应电荷
induced current 感生电流,感应电流
induced electric field 感生电场,感应电场
induced e.m.f. 感生电动势,感应电动势
induced emission 感生发射,受迫发射
induced fission 诱发裂变
induced giant moment 感生巨磁矩
induced moment 感生力矩
inductance 电感,感应系数
induction 感应,感应现象
induction coil 感应(线)圈,电感线圈
induction field 感应场
induction heating 感应加热
induction motor 感应电动机
inductive coupling 电感耦合
inductive reactance 感抗
inductivity 感应率
inductor 电感器,感应体,感应器
inelastic collision 非弹性碰撞
inelasticity 非弹性;非弹性系数,非弹性度
inelastic scattering 非弹性散射
inelastic scattering cross-section 非弹性散射截面
inequality 不等式;不等
inertia 惯性,惯量
inertial coefficient 惯性系数
inertial confinement 惯性约束
inertial force 惯性力
inertial frame 惯性坐标系
inertial mass 惯性质量
inertial system 惯性系
inertia tensor 惯量张量,惯性张量
infinite series 〈数〉无穷级数
infinitesimal 无穷小的,无限

小的
infinity 无穷大，无限
inflation 膨胀
inflationary universe 暴胀宇宙
inflection 凹陷，拐折
inflection point 〈数〉拐点
information capacity 信息容量
information channel 信道
information entropy 信息熵
information optics 信息光学
infranics 红外电子学
infrared astronomy 红外天文学
infrared divergence 红外发散
infrared lamp 红外灯
infrared laser 红外激光器
infrared radiation (IR) 红外辐射
infrared ray 红外线
infrared spectrophotometer 红外分光光度计
infrared spectroscopy 红外光谱学
infrared stimulated luminescence 红外受激发光
infrared thermometer 红外线温度计
infrasonic region 次声区
infrasonic wave 次声波
infrasound 次声
initial condition 初始条件
initial state 初态
initial velocity 初速(度)
injection laser 注入式激光器
in-phase 同相
input device 〈计算机〉输入设备
insolation 日照辐射
instability 不稳定性
instantaneous axis of rotation 转动瞬轴，瞬时转动轴，转动即时轴
instantaneous centre 瞬心，瞬时中心，即时中心
instantaneous value 瞬时值
instantaneous velocity 瞬时速度，即时速度
instanton 瞬子
instrument 仪器
instrumental optics 仪器光学
insulant 绝缘材料
insulated-gate field effect transistor (IGFET) 绝缘栅场效(应)管
insulating medium 绝缘介质
insulation 绝缘
insulator 绝缘体，绝热器
integrable system 可积系统
integral calculus 〈数〉积分学
integrand 〈数〉被积函数
integrated circuit (IC) 集成电路
integrated optics 集成光学
integrated optical circuit 集成光路
integration 集成；〈数〉积分法
intensity 强度，光亮度
intensive quantity 强度量
intensive variable 强度变量
interaction 交互作用；相互作用
interatomic force 原子间力
interband transition 带间跃迁
intercalation 插层，置闰
interface 界面
interface state 界面态
interference 干涉，干扰
interference colour 干涉色
interference filter 干涉滤光片，干涉滤波器
interference fringe 干涉条纹
interference microscope 干涉显微镜
interference pattern 干涉图样
interference term 干涉项
interference wave 干涉波
interferometer 干涉仪，干扰计
interferometry 干涉测量术，干涉量度学
intergalactic space 〈天文〉星系际空间
intergrowth 共生，交互生长
interlayer 界层，隔层
intermediate bond 居间键
intermediate coupling 中等耦合，居间耦合
intermediate frequency (IF) 中频

intermediate image 居间像，中间图像
intermediate neutron 中能中子
intermediate phase 〈冶金〉中间相，居间相
intermediate state 居间态，中间态
intermediate vector boson (IVB) 中间矢量玻色子
intermetallic compound 〈化〉金属互化物
intermolecular force 分子间力
internal combustion engine 内燃机
internal conversion 内转换，内部变换
internal conversion coefficient 内转换系数
internal energy 内能
internal force 内力
internal friction 内摩擦，内耗
internal photoelectric effect 内光电效应
internal reflection 内反射
internal resistance 内阻，内电阻
internal symmetry 内部对称性
international candle 国际烛光
international date line 国际日期变更线，国际日界线
International Practical Temperature Scale 国际实用温标
international symbol 国际符号
International Union of Pure and Applied Physics (IUPAP) 国际纯粹物理与应用物理联合会
Internet （国际）互联网
interplanetary space 行星际太空，行星际空间
interpolation 内插
interpolation function 插值函数
interpolation polynomial 内插多项式，插值多项式
interstellar gas 星际气体
interstellar space 星际空间
interstitial 填隙原子，填隙子
interstitial compound 填隙式化合物
interval of events 事件间隔
intonation 声调
intrinsic angular momentum 内禀角动量
intrinsic carrier 内禀载流子，本征载流子
intrinsic energy 内能，本征能
intrinsic equation 内禀方程，禀性方程，本征方程
intrinsic semiconductor 本征半导体，内禀半导体
introscope 内窥镜
Invar 不胀钢，殷钢
inverse Compton effect 逆康普顿效应
inverse Compton scattering 逆康普顿散射
inverse function 〈数〉反函数，逆函数
inverse magnetooptical effect 逆磁光效应
inverse operation 逆操作
inverse-square law 平方反比定律
inversion 反演
inversion layer 反型层，逆温层
inversion temperature 反转温度
inverted image 倒像
inverter 反相器，倒相器
inverting prism 倒像棱镜，转像棱镜
involute 渐开线，切展线，渐伸线
ion 离子
ion beam 离子束
ion-beam analysis (IBA) 离子束分析
ion-beam generator 离子束发生器
ion bombardment 离子轰击
ion engine 离子发动机
ion exchange 离子交换
ion exchange adsorption 离子交换吸附
ionic bond 离子键

ionic bonding 离子键合
ionic crystal 离子晶体
ionic radius 离子半径
ionic strength 离子强度
ion implantation 离子注入,离子植入
ionization 离子化,电离
ionization by collision 碰撞电离
ionization chamber 电离室,电离箱
ionization current 电离电流
ionization energy 电离能
ionization gauge 电离压力计,电离压强计;电离真空规,离子规
ionization potential (IP) 电离电势,电离电位
ionization spectroscopy 电离谱学
ionizing radiation 电离辐射
ion-microprobe analysis 离子微探针分析
ionosphere 电离层
ionospheric wave 电离层波
ion pair 离子偶,离子对
ion pump 离子泵
ion scattering spectroscopy 离子散射谱学
IP (ionization potential) 电离电势,电离电位
IR (infrared radiation) 红外辐射
iris 可变光阑;虹膜
irradiance 辐照度
irradiation 光渗,辐照
irradiation damage 辐照损伤
irradiation effect 辐照效应
irrational number 〈数〉无理数
irregular reflection 不规则反射
irrelevant parameter 无关参量
irreversibility 不可逆性
irreversible process 不可逆过程
irrotational flow model 无旋流模型
isentropic compression 等熵压缩
isobar 同量异位素;等压线
isobaric nucleus 同质异位核
isobaric spin 同位旋
isochore 等容线,等体积线
isochoric process 等容过程,等体积过程
isochronism 等时性
isochronous oscillation 等时振荡
isoclinal fold 等斜褶皱
isocline 等斜线
isoclinic (line) 等磁倾线,等倾线
isodiaphere 等超额中子核素
isodynamic line 等磁力线
isoelectric point 等电点,等电离点
isoelectronic 等电子的
isoelectronic group 等电子群
isogonal line 等偏线
isogyre 同消色线
isolated conductor 孤立导体
isolated system 孤立系
isolation 隔离;绝缘
isolation table 防震台
isolator 绝缘体
isomagnetic line 等磁线
isomerism 同质异能性;同分异构性
isomer 同核异能素,同质异能素;同分异构体
isometric 等轴的,立方的
isometric system 立方系,立方晶系
isomorphism 同晶性,同形性,类质同晶
isopach 等厚线
isophote 等强度线,等照度线,等辐透线
isoplanatic region 等晕区
isopotential line 等势线
isopotential surface 等势面,等电位面
isospin 同位旋
isotherm 等温线
isothermal process 等温过程
isotone 同中子异位素,等中子异位素
isotonic 等渗压的
isotope 同位素
isotope effect 同位素效应
isotope separation 同位素分离

isotope shift 同位素移位
isotopic abundance 同位素丰度
isotopic number 中(子)质(子数)差
isotopic spin 同位旋
isotopy 同位素学
isotron 同位素分析器,同位素分离器
isotropic medium 各向同性介质
isotropism 各向同性
isotropy 各向同性
iteration method 迭代法
iterative process 迭代过程
itinerant electron 巡游电子
IUPAP (International Union of Pure and Applied Physics) 国际纯粹物理与应用物理联合会
IVB (intermediate vector boson) 中间矢量玻色子

Jj

J (joule) 焦耳(功、能单位)
jack 插孔,插座;千斤顶
Jacobian (determinant)〈数〉雅可比行列式
jellium model 凝胶模型
jet 喷嘴,喷射器;射流;喷气;喷注
jet electrification 喷注起电
jet engine 喷气发动机
jet model 喷注模型
jet plane 喷气式飞机
jet propeller 喷射推进器,喷气式螺旋桨
jet propulsion 喷气推进
jet pump 喷注泵,喷注抽机
jet shower 喷注簇射
jet stream 急流,射流,喷流
JFET (junction field effect transistor) 结型场效(应)管
j-j coupling j-j 耦合
Josephson device 约瑟夫森器件
Josephson effect 约瑟夫森效应
joule (J) 焦耳(功、能单位)
Joule experiment 焦耳实验
Joule heat 焦耳热
Joule heating 焦耳加热
Joule-Kelvin effect 焦耳-开尔文效应
Joule's laws 焦耳定律
Joule-Thomson coefficient 焦耳-汤姆孙系数
Joule-Thomson effect 焦耳-汤姆孙效应
Jovian planet 类木行星
J/psi particle J/ψ 粒子
JUGFET (junction gate field effect transistor) 结栅场效应晶体管
jump 突跳,跳变;跃迁
junction detector 结型检波器
junction field effect transistor (JFET) 结型场效(应)管
junction gate field effect transistor (JUGFET) 结栅场效应晶体管
junction transistor 结型晶体管,面结型晶体管

Kk

K (kelvin) 开(温度单位)
Kaluza-Klein theory 卡鲁扎-克莱因理论
KAM theorem (Kolmogorov-Arnold-Moser theorem) KAM 定理
KAM torus KAM 环面
kaon K 介子
karat 克拉(宝石重量单位);开金
Kater's pendulum 凯特摆
katharometer 热导计,导热析气计
keeper 卫铁
kelvin (K) 开(温度单位)
Kelvin effect 开尔文效应,趋肤效应
Kelvin (temperature) scale 开氏温标,开尔文温标,绝对温标
Kepler's laws 开普勒定律
kerma 比释动能
Kerr effect 克尔效应
Kerr solution 克尔解
ket 右矢,刃
kg (kilogramme) 千克,公斤(质量单位)
kHz (kilohertz) 千赫(频率单位)
kilocalorie 大卡,千卡(热量单位)
kilogramme (kg) 千克,公斤(质量单位)
kilogramme calorie 大卡,千卡(热量单位)
kilohertz (kHz) 千赫(频率单位)
kilometre (km) 千米,公里(长度单位)
kilowatt (kW) 千瓦(功率单位)
kilowatt-hour (kWh) 千瓦时,度(电能单位)
kinematic equation 运动学方程
kinematics 运动学
kinematic viscosity 运动黏度
kinetic energy 动能
kinetic equation 动力学方程,动理(学)方程
kinetic friction 动摩擦
kinetic mode 动理模
kinetic potential 动理势
kinetics 动力学,动理学
kinetic theory 分子动理论,分子运动论
kinetic theory of gases 气体动理论,气体分子运动论
kinetic viscosity 动力黏度
Kirchhoff's law of radiation 基尔霍夫辐射定律
Kirchhoff's laws 基尔霍夫定律
Klein-Gordon equation 克莱因-戈尔登方程
klystron (电子)速调管
km (kilometre) 千米,公里(长度单位)
knocking 爆震音;爆击
Knudsen effect 克努森效应
Knudsen flow 克努森流,过渡流
Kohlrausch square root law 科尔劳希平方根定律
Kolmogorov-Arnold-Moser theorem (KAM theorem) KAM 定理
Kovar 科瓦铁镍钴合金
Kramers theorem 克拉默斯定理
Kundt's tube 孔特管
kW (kilowatt) 千瓦(功率单位)
kWh (kilowatt-hour) 千瓦时,度(电能单位)

Ll

L; l (litre) 升(体积单位)
labelling 标记;示踪
laboratory scale 实验室尺度
ladder approximation 梯图近似,阶梯近似
ladder diagram 梯图
lag 移后;落后;滞后
Lagrangian equation 拉格朗日方程
Lagrangian function 拉格朗日函数
lambda particle λ粒子
lambda point λ点
Lamb dip 兰姆凹陷
lambert 朗伯(亮度单位)
Lambert's laws 朗伯定律
Lamb shift 兰姆移位
laminar flow 层流
laminated core 叠片铁芯
Landau damping 朗道阻尼
Landau level 朗道能级
Landé g-factor 朗德g因子
Landé interval rule 朗德间隔定则
Langevin equation 朗之万方程
Langevin function 朗之万函数
Langmuir frequency 朗缪尔频率
Langmuir oscillation 朗缪尔振荡
Langmuir probe 朗缪尔探针
Langmuir wave 朗缪尔波
Laplace equation 拉普拉斯方程
large-scale integrated circuit (LSI circuit) 大规模集成电路
Larmor frequency 拉莫尔频率
Larmor precession 拉莫尔进动,拉莫尔旋进
Larmor radius 拉莫尔半径
laser 激光;激光器
laser aligning 激光校直
laser amplifier 激光放大器
laser annealing 激光退火
laser array 激光阵列
laser beam 激光束
laser bonding 激光焊接
laser boring 激光打孔
laser cavity 激光(共振)腔
laser cooling 激光制冷
laser cutting 激光切割
laser diode 激光二极管
laser excitation 激光激发
laser guidance 激光制导
laser gyroscope 激光陀螺
laser host 激光(器)基质
laser oscillator 激光振荡器
laser physics 激光物理学
laser power meter 激光功率计
laser printer 激光打印机
laser pump 激光泵
laser pumping 激光抽运
laser range finder 激光测距仪
laser spectrum 激光光谱
latent heat 潜热
latent heat of fusion 熔化潜热
latent heat of vaporization 汽化潜热
latent image 潜像
lateral aberration 横向像差
lateral coherence 横向相干性
lateral inversion (图像)左右倒置,横向倒置
lateral magnification 横向放大率
lateral velocity 横向速度
latitude 纬度,纬线
lattice 格点;点阵;晶格

lattice constant 晶格常量
lattice dynamics 晶格动力学
lattice energy 晶格能,点阵能
lattice gauge theory 格点规范理论
lattice heat capacity 晶格热容
lattice mismatch 晶格失配
lattice point 格点
lattice scattering 晶格散射
lattice site 格座,点阵座
lattice theory 晶格理论,点阵理论
lattice vector 格矢,点阵矢量
lattice vibration 晶格振动,点阵振动
lattice vibrational absorption 晶格振动吸收
lattice wave 格波
lattice wave momentum 格波动量
Laue diffraction 劳厄衍射
Laue method 劳厄法
Laue-symmetry 劳厄对称
Laue-symmetry group 劳厄对称群
law 定律
law of conservation of angular momentum 角动量守恒定律
law of conservation of charge 电荷守恒定律
law of conservation of energy 能量守恒定律,能量守恒与转化定律
law of conservation of mass 质量守恒定律
law of conservation of mechanical energy 机械能守恒定律
law of conservation of moment of momentum 动量矩守恒定律
law of conservation of momentum 动量守恒定律
law of conservation of parity 宇称守恒定律
law of inertia 惯性定律
law of reflection 反射定律
law of refraction 折射定律
law of symmetry 对称定律
law of universal gravitation 万有引力定律
layered ultrathin coherent structure (LUCS) 多层超薄共格结构
layer structure 层状结构
LCD (liquid crystal display) 液晶显示器
LDR (light-dependent resistor) 光敏电阻(器)
lead-acid accumulator 铅酸蓄电池
lead-acid cell 铅酸电池
leading particle 主导粒子
leading pole 主导极点,导磁极
leading term 主导项
leading wave 前导波
leakage current 漏泄电流
leakage factor 漏泄因子,渗漏系数
leakage flux 漏磁通,漏通量
leakage resistance 漏电阻,泄漏电阻
least-squares method 最小二乘法
Leclanché cell 勒克朗谢电池
LED (light-emitting diode) 发光二极管
LEED (low-energy electron diffraction) 低能电子衍射
Lee model 李(政道)模型
left-hand circular polarization 左旋圆偏振
left-handed crystal 左旋晶体
left-hand rule 左手定则
Legendre transformation 勒让德变换
length 长度
length contraction 长度收缩
lens 透镜,镜头
lens centre 透镜中心,透镜光心
lens combination 透镜组
lens formula 透镜公式
lensless Fourier hologram 无透镜傅里叶全息图

lens spectrometer 透镜谱仪,透镜分光计
Lenz's law 楞次定律
lepton 轻子
leptonic decay 轻子型衰变
Leslie's cube 来斯利立方体
LET (linear energy transfer) 线能量转移,线性能量转移
level 能级;水准器;水位;电平
level-crossing effect 能级交叉效应
level density 能级密度,电平密度
lever 杆;杠杆;手柄
lever arm 杠杆臂
lever rule 杠杆定则
levorotation 左旋;左旋光
Leyden jar 莱顿瓶
LF (low frequency) 低频
l-form 左旋形式
libration 天平动
lidar (light detection and ranging) 激光雷达
lifetime 寿命
lift force 升力
ligand field theory 配体场论
light 光
light amplification 光放大,光增益
light amplifier 光增强器
light beam 光束,光柱
light beat 光拍
light bulb 灯泡
light chopper 斩光器,截光器,遮光器
light cone 光锥
light-dependent resistor (LDR) 光敏电阻(器)
light detection and ranging (lidar) 激光雷达
light emission 光发射
light-emitting diode (LED) 发光二极管
light energy 光能
light field 光场
light guide 光导
lighting device 照明装置
lightning 闪电
lightning conductor 避雷装置,避雷线
lightning rod 避雷针
light nucleus 轻核
light pencil 光束,光锥
light pressure 光压
light pulse 光脉冲
light pulse compression 光脉冲压缩
light pump 光泵
light quantum 光量子
light ray 光线
light scattering 光散射
light-sensitive cathode 光敏阴极
light sensor 光敏传感器,光敏元件
light shift 光致频移
light source 光源
light trap 光阱,陷光器
light valve 光阀
light velocity 光速
light wave 光波
light year 光年(长度单位)
limit 〈数〉极限,限度
limit cycle 极限环
limiting group 极限群
limiting velocity 极限速度
linac 直线加速器,线性加速器
Linde cycle 林德循环
Linde liquefier 林德液化机
linear absorption coefficient 线性吸收系数
linear accelerator 直线加速器,线性加速器
linear amplifier 线性放大器
linear charge density 线电荷密度
linear density 线密度
linear element 线性元件
linear energy transfer (LET) 线能量转移,线性能量转移
linear equation 〈数〉线性方程
linear expansivity 线膨胀率
linear magnification 线性放大;线性放大率
linear motor 直线电动机

linear optics 线性光学
linear pinch 直线箍缩
linear polarization 线偏振
linear resolution 线分辨率
linear space 线性空间
line breadth 谱线宽度,线宽
line broadening 谱线增宽
line defect 线缺陷
line element 线元,线素
line integral 〈数〉线积分
line of sight 视线,瞄准线
line printer 行式打印机
line profile 谱线轮廓
line series 线系,谱线系
line shape 线形,谱线形状
lines of force 力线
line spectrum 线状谱,线光谱
line strength 谱线强度
line width 谱线宽度,线宽
liquation 熔化分离作用
liquefaction 液化
liquefaction point 液化点
liquefied petroleum gas (LPG) 液化石油气
liquefier 液化器
liquid 液体
liquid crystal 液晶
liquid crystal display (LCD) 液晶显示器
liquid crystal phase 液晶相
liquid-drop model 液滴模型
liquid helium 液氦
liquid-in-glass thermometer 玻璃液体温度计
liquid laser 液体激光器
liquid nitrogen 液氮
liquidus (line) 液相线
Lissajous figures 利萨如图形
lithium battery 锂电池
litre (L; l) 升(体积单位)
Lloyd's mirror 劳埃德镜
lm (lumen) 流明(光通量单位)
load 负载,载荷
Local Group 〈天文〉本星系群
localization 定域化;局部化;定位
local oscillator 本机振荡器
local theory 局部理论,定域理论
locus 轨迹
lodestone 天然磁石
logarithm 对数
logarithmic scale 对数标度,对数比例,对数标尺
logarithmic series 对数级数
logic circuit 逻辑电路
lone-pair electron 孤对电子
longitude 经度,经线
longitudinal aberration 纵向像差
longitudinal magnification 纵向放大率,轴向放大率
longitudinal mass 纵质量
longitudinal mode 纵模
longitudinal relaxation 纵向弛豫
longitudinal wave 纵波
long-range force 长程力
long-sightedness 远视
long wave 长波
loop 回路;循环;环
Lorentz force 洛伦兹力
Lorentz group 洛伦兹群
Lorentz-Lorenz equation 洛伦兹-洛伦茨方程
Lorentz transformations 洛伦兹变换
Loschmidt number 洛施密特常量
Loschmidt's constant 洛施密特常量
loudness 响度,音量
loudspeaker 扬声器,喇叭,扩音器
low dimensional physics 低维物理学
low dimensional solid 低维固体
low-energy electron diffraction (LEED) 低能电子衍射
lowering of vapour pressure 蒸汽压降低
lowest unoccupied molecular orbit (LUMO) 最低未占分子轨道
low frequency (LF) 低频
low temperature physics 低温物理学

low-temperature plasma 低温等离子体
low-temperature thermometer 低温计
LPG (**liquefied petroleum gas**) 液化石油气
L-S coupling L-S耦合,罗素-桑德斯耦合
LSI circuit (**large-scale integrated circuit**) 大规模集成电路
lubrication 润滑,润滑作用
LUCS (**layered ultrathin coherent structure**) 多层超薄共格结构
lumen (**lm**) 流明(光通量单位)
luminance 亮度
luminescence 发光;发光学
luminiferous ether 光以太
luminometer 照度计
luminophor 发光体
luminosity 发光度,亮度
luminous exitance 光出射度,光束发散度
luminous flux 光通量
luminous intensity 发光强度
LUMO (**lowest unoccupied molecular orbit**) 最低未占分子轨道
lunar eclipse 〈天文〉月食
lunation 〈天文〉朔望月,太阴月
lux (**lx**) 勒,勒克斯(照明单位)
luxmeter 照度计,勒克斯计
lx (**lux**) 勒,勒克斯(照明单位)
Lyman series 莱曼系
lyoluminescence 晶溶发光,水合发光
lyotropic liquid crystal 溶致液晶

Mm

m (metre) 米(长度单位)
MA (mechanical advantage) 机械利益
Machmeter 马赫计,马赫表
Mach number 马赫数
Mach's principle 马赫原理
Maclaurin's series 〈数〉麦克劳林级数
macrocausality 宏观因果性
macrocrystal 宏晶,粗晶,大块结晶
macromolecule 高分子,大分子
macroscopic 宏观的
macroscopic polarization 宏观极化
macroscopic quantity 宏观量
macroscopic quantum phenomenon 宏观量子现象
macroscopic state 宏观态
macrostructure 宏观结构
Magellanic clouds 麦哲伦云
magic nucleus 幻核
magic number 幻数
maglev 磁悬浮列车
maglev train 磁悬浮列车
Magnalium 镁铝合金
magnet 磁体,磁铁
magnetic aftereffect 磁后效
magnetically levitated train 磁悬浮列车
magnetic anisotropy 磁各向异性
magnetic axis 磁轴
magnetic bottle (等离子)磁瓶
magnetic bubble 磁泡
magnetic bubble memory 磁泡存储器
magnetic charge 磁荷
magnetic circuit 磁路
magnetic compass 罗盘
magnetic confinement (等离子)磁约束
magnetic constant 磁常量,磁性常数
magnetic declination 磁偏角
magnetic deflection 磁偏转
magnetic dip 磁倾角
magnetic dipole 磁偶极子
magnetic dipole moment 磁偶极矩
magnetic dipole radiation 磁偶极辐射
magnetic disk 〈计算机〉磁盘
magnetic domain 磁畴
magnetic elements 地磁要素
magnetic energy 磁能
magnetic energy density 磁能密度
magnetic energy product 磁能积
magnetic equator 地磁赤道
magnetic field 磁场
magnetic field intensity 磁场强度
magnetic field line 磁场线
magnetic field strength 磁场强度
magnetic fluid 磁(性)流体
magnetic flux 磁通量
magnetic flux density 磁通密度,磁感应强度
magnetic focusing 磁聚焦
magnetic force 磁力
magnetic group 磁群
magnetic hysteresis 磁滞现象
magnetic hysteresis loop 磁滞回线

magnetic hysteresis loss 磁滞损耗
magnetic induction 磁感应强度
magnetic intensity 磁场强度
magnetic lens 磁透镜
magnetic material 磁性材料
magnetic medium 磁介质
magnetic meridian 磁子午圈
magnetic mirror 磁镜
magnetic moment 磁矩
magnetic monopole 磁单极子
magnetic needle 磁针,指南针
magnetic north 磁北
magnetic opticity 磁致旋光
magnetic permeability 磁导率
magnetic pole 磁极
magnetic potential 磁位
magnetic probe 磁探针
magnetic pump 磁力泵
magnetic pumping 磁抽运
magnetic quantum number 磁量子数
magnetic recording 磁记录
magnetic refrigeration 磁致冷
magnetic relaxation 磁弛豫
magnetic reluctance 磁阻
magnetic resistance 磁阻
magnetic resonance 磁共振
magnetic rigidity 磁刚度,磁硬度
magnetic rotation 磁致旋光
magnetics 磁学
magnetic semiconductor 磁性半导体
magnetic shell 磁壳
magnetic shielding 磁屏蔽
magnetic spectrometer 磁谱仪
magnetic spectrum 磁谱
magnetic storage 磁存储
magnetic storm 磁暴
magnetic surface 磁面
magnetic susceptibility 磁化率
magnetic symmetry 磁对称性
magnetic tape 磁带
magnetic tape recorder 磁带录音机
magnetic transition 磁跃迁,磁转变
magnetic tuning 磁调谐
magnetic tuning laser 磁调谐激光器
magnetic variation 磁变
magnetic viscosity 磁黏(滞)性
magnetic viscosity coefficient 磁黏性系数
magnetism 磁性,磁学
magnetization 磁化,起磁,磁化强度
magnetization curve 磁化曲线
magnetizer 磁化器,磁化机
magneto 磁电机,永磁发电机
magnetoacoustic effect 磁声效应
magnetoacoustic transfer device 磁声转换器件
magnetobremsstrahlung 磁韧致辐射
magnetocaloric effect 磁致变温效应,磁热效应
magnetochemistry 磁化学
magnetocrystalline 磁晶
magnetoelastic effect 磁弹效应
magnetoelasticity 磁致弹性
magnetoelectric effect 磁电效应
magnetoelectricity 磁电,磁电学
magnetogasdynamics 磁性气体动力学
magnetograph 地磁记录仪,磁强记录仪
magnetohydrodynamics (MHD) 磁流体动力学
magnetomechanical effect 磁-力效应
magnetomechanical ratio 磁机工比率
magnetometer 磁强计,地磁仪
magnetomotive force (m. m. f.) 磁动势,磁通势
magneton 磁子
magneto-optical effect 磁光效应
magneto-optics 磁光学
magnetopause 磁层顶,磁顶
magnetophone 磁带录音机
magnetoresistance 磁致电阻

magnetoresistance effect 磁致电阻效应,磁阻效应
magnetoscope 验磁器
magnetosensitive sensor 磁敏器件
magnetosonic wave 磁声波
magnetosphere 磁层,磁气圈
magnetostatics 静磁学
magnetostriction 磁致伸缩
magnetron 磁控管
magnetron sputtering 磁控溅射
magnification 放大;放大率
magnifier 放大镜
magnifying glass 放大镜
magnifying power 放大率
magnitude 量级,〈天文〉星等
magnitude difference 〈天文〉星等差
magnon 磁波子,磁振子
Magnox 镁诺克斯合金
Magnus effect 马格努斯效应
mainframe computer 大型计算机
main-sequence star 〈天文〉主序星
main shaft 主轴
majority carrier 多(数载流)子
majority carrier density 多子浓度
majority rule 多数规则
Maksutov telescope 马克苏托夫望远镜
Maltese cross 马尔特十字接头
Malus law 马吕斯定律
manganin 锰铜,锰镍铜合金
manometer 流体压强计,气压计
mantissa 〈数〉(对数的)尾数
many-body problem 多体问题
Mariotte's law 马略奥特定律
Markovian process 马尔可夫过程
Markov process 马尔可夫过程
martensite 〈冶金〉马氏体,马顿斯体
mascon 质量密集物质
maser 微波激射;微波激射器
mask 掩模;掩蔽,屏蔽
mass 质量
mass action 质量作用
mass action law 质量作用(定)律
mass decrement 质量减损
mass defect 质量亏损
mass-energy equivalence 质能等价性
mass-energy relation 质能关系
mass excess 质量过剩,质量盈余
mass formula 质量公式
mass number 质量数,核子数
mass-number conservation 质量数守恒
mass point 质点
mass spectrometer 质谱仪,质谱分析器
mass spectrum 质谱
mass stopping power 质量阻止本领
mass transfer 传质,质量转移
master equation 主方程
master routine 主程序
matched filter 匹配滤波器
material point 质点
material science 材料科学
mathematical pendulum 数学摆,单摆
mathematical physics 数理物理学
matrix 〈数〉矩阵
matrix mechanics 矩阵力学
matter 物质
matter wave 物质波
maximum deviation 最大偏差
maximum error 最大误差
maximum likelihood method 最大似然法
maximum magnetic energy product 最大磁能积
maximum permissible dose 最大容许剂量
maximum thermometer 最高温度表
maxwell 麦,麦克斯韦(磁通量单位)

Maxwell-Boltzmann distribution 麦克斯韦-玻尔兹曼分布
Maxwell-Boltzmann statistics 麦克斯韦-玻尔兹曼统计法
Maxwell relation 麦克斯韦关系
Maxwell's demon 麦克斯韦妖
Maxwell's equations 麦克斯韦方程组
Maxwell speed distribution 麦克斯韦速率分布
Maxwell stress tensor 电磁场应力张量
Maxwell velocity distribution 麦克斯韦速度分布
Mayer function 迈耶函数
mB (millibar) 毫巴(压强单位)
MBE (molecular beam epitaxy) 分子束外延
mC (millicurie) 毫居里(放射性活度单位)
McLeod gauge 麦克劳德压力计,压缩式真空计
McLeod vacuum gauge 麦克劳德真空规
Md (mendelevium) 〈化〉钔
mean 平均数;中数;平均的
mean anomaly 〈天文〉平近点角
mean deviation 平均偏差
mean error 平均误差
mean free path 平均自由程
mean free time 平均自由时间
mean lifetime 平均寿命
mean range 平均射程
mean solar day 〈天文〉平太阳日
mean speed 平均速率
mean square displacement 均方位移
mean velocity 平均速度
measurement 测量,量度
measurement error 测量误差
mechanical advantage (MA) 机械利益
mechanical energy 机械能
mechanical equilibrium condition 力学平衡条件
mechanical equivalent of heat 热功当量
mechanical equivalent of light 光功当量
mechanical mass 力学质量
mechanical motion 机械运动,力学运动
mechanical quality factor 力学品质因数
mechanical system 力学系(统),机械系
mechanical vibration 机械振动
mechanical wave 机械波
mechanical work 机械功
mechanics 力学
mechanism 机械装置;机理,结构
mechanocaloric effect 力热效应,机械致热效应
median 中线;中值
medium 介质,媒质
medium frequency (MF) 中频
medium wave (mw) 中波
megahertz (MHz) 兆赫(频率单位)
megohmmeter 兆欧计
Meissner effect 迈斯纳效应
melt 熔化,熔融;熔体,熔化物
melting heat 熔化热
melting point (m.p.) 熔点
membrane 薄膜,膜片,隔膜
mendelevium (Md) 〈化〉钔
meniscus 弯月形透镜;弯月面
Mercury 〈天文〉水星
mercury 〈化〉汞,水银
mercury barometer 水银气压计
mercury cell 汞极电池
mercury column 汞柱,水银柱
mercury lamp 汞灯,水银灯
mercury-vapour lamp 汞汽灯
meridian 子午线
meson 介子
mesophase 中介相
mesoscopic physics 介观物理学
mesoscopic structure 介观结构
metagalaxy 〈天文〉总星系
metal 金属
metal fatigue 金属疲劳

metal-insulator-semiconductor structure 金属-绝缘体-半导体结构
metal-insulator transition 金属-绝缘体转变
metallic bonding 金属键合
metallic ceramics 金属陶瓷
metallic glass 金属玻璃
metallic lattice 金属光栅，金属点阵
metallography 金属组织学，金相学
metallurgy 冶金学，冶金术
metal-oxide-semiconductor field effect transistor（MOSFET）金属-氧化物-半导体场效应晶体管，MOS 场效管
metal-oxide-semiconductor structure 金属-氧化物-半导体结构
metal physics 金属物理学
metal-semiconductor contact 金属-半导体接触
metastability 亚稳定性
metastable equilibrium 亚稳平衡
metastable phase 亚稳相
metastable state 亚稳态
meteor 〈天文〉流星
meteorite 〈天文〉陨星
meteoroid 〈天文〉宇宙尘，流星体
meter 测量仪表，计
metglass 金属玻璃
method of images 镜像法，电像法
method of mixtures 混合法
method of partial waves 分波法
method of pseudopotential 赝势法
metre（m）米（长度单位）
metre bridge 滑线电桥
metre rule 米尺
metric system 米制
metric tensor 度规张量
metric ton 公吨（质量单位）
metrology 度量衡学，计量学；度量衡制
MF（medium frequency）中频
mg（milligram）毫克（质量单位）
MHD（magnetohydrodynamics）磁流体动力学
MHz（megahertz）兆赫（频率单位）
mho 姆欧（电导率单位）
Michelson interferometer 迈克耳逊干涉仪
Michelson-Morley experiment 迈克耳逊-莫雷实验
Michelson stellar interferometer 迈克耳逊测星干涉仪
microampere（μA）微安（电流强度单位）
microbalance 微量天平
microcosm 微观宇宙
microcrystal 微晶（体）
microcurie（μCi）微居里（放射性活度单位）
microdefect 微缺陷
microdensitometer 测微光度计，测微密度计
microelectronics 微电子学
microfilm 缩微胶片
microimage storage 缩微像存储
micrometeorite 〈天文〉微小陨石，微陨星
micrometer 测微计，千分尺
micrometer caliper 螺旋测微器，螺旋测径器，千分测径规
micrometer eyepiece 测微目镜
micron（μm）微米（长度单位）
microphone 话筒，麦克风，扩音器
microphotograph 缩微照片，显微照片
microphotometer 显微光度计，测微光度计
microphysics 微观物理学
microprocessor 〈计算机〉微处理器
microscope 显微镜
microscopic 微观的，显微镜的
microscopic particle 微观粒子
microscopic quantity 微观量

microscopic state 微观态
microspectrophotometer 测微分光光度计
microstructure 微结构,显微结构
microtron 电子回旋加速器
microvolt 微伏(电压单位)
microwave 微波
microwave background radiation 微波背景辐射
microwave optics 微波光学
microworld 微观世界
mile 英里(长度单位)
Milky Way 银河系
Milky Way Galaxy 银河系
Miller indices 米勒指数
milliammeter 毫安表,毫安计
millibar (mB) 毫巴(压强单位)
millicurie (mC) 毫居里(放射性活度单位)
milligram (mg) 毫克(质量单位)
Millikan oil-drop experiment 密立根油滴实验
millilitre (ml) 毫升(容积单位)
millimetre (mm) 毫米(长度单位)
minimum thermometer 最低温度计
Minkowski geometry 闵可夫斯基几何
Minkowski space 闵可夫斯基空间
minority carrier 少(数载流)子
minority carrier density 少子浓度
minor planet 〈天文〉小行星
minute 分钟(时间单位)
mirage 海市蜃楼,蜃景
mirror 镜子,反射镜
mirror image 镜像,反像
mirror nuclei 镜像核
mirror reflection 镜(面)反射,镜像反射
missing mass 短缺质量,丢失质量
missing order 缺级,缺序
mixed crystal 混晶
mixed state 混合态
mixer 混频器,混合器
mixing 混合性,混频
m. k. s. units 米·千克·秒制
ml (millilitre) 毫升(容积单位)
mm (millimetre) 毫米(长度单位)
m. m. f. (magnetomotive force) 磁动势,磁通势
mmHg 毫米汞柱(压强单位)
mobile defect 可动缺陷
mobile equilibrium 可动平衡,动态平衡
mobility 迁移率,动性
mobility edge 迁移率边
mobility gap 迁移率隙
modality 模态
mode 模,波型
mode configuration 模结构
mode coupling 模耦合,波型耦合
mode degeneracy 波模简并度
model 模型
mode-locking 锁模,波模锁定,模式锁定
modem (modulator-demodulator) 调制解调器
mode number 模数
mode of laser 激光模(式)
mode of vibration 振动模(式)
mode pattern 模图样,波型图样
mode pulling effect 模牵引效应
mode pushing effect 模推斥效应
moderator 减速剂,缓和剂
modern physics 近代物理学
mode selection 选模
mode sequence 模序列
mode suppression 模抑制
modulation 调制
modulation spectroscopy 调制光谱学
modulation transfer function (MTF) 调制传递函数
modulator-demodulator (modem) 调制解调器
modulus 模数,模量
modulus of elasticity 弹性模量,

弹性模数
Moh hardness scale 莫氏硬度标
Mohorovičić discontinuity 莫霍罗维奇界面
moiré fringe 叠栅条纹,莫尔条纹
moiré pattern 云纹图
molar 摩尔的
molar conductivity 摩尔导电率
molar heat capacity 摩尔热容
molar volume 摩尔体积
mole 摩尔
molecular beam 分子束
molecular beam epitaxy (MBE) 分子束外延
molecular crystal 分子晶体
molecular current 分子电流
molecular distillation 分子蒸馏
molecular electronics 分子电子学
molecular field theory 分子场理论
molecular flow 分子流
molecular formula 〈化〉分子式
molecular hypothesis 分子假说
molecular magnet 分子磁体
molecular orbital 分子轨函数
molecular physics 分子物理学
molecular scattering 分子散射
molecular sieve 分子筛
molecular solid 分子固体
molecular spectroscopy 分子光谱学
molecular spectrum 分子光谱
molecular structure 分子结构
molecular theory 分子说,分子理论
molecular volume 摩尔体积,分子体积
molecular weight 分子量
molecule 分子
mole fraction 摩尔分数
moment 力矩,矩
moment of couple 力偶矩
moment of force 力矩
moment of inertia 转动惯量
moment of momentum 动量矩,角动量
momentum 动量
momentum conservation 动量守恒
momentum representation 动量表象
momentum space 动量空间
monoatomic gas 单原子气体
monochromaticity 单色性
monochromatic light 单色光
monochromatic radiation 单色辐射
monochromatic source 单色光源
monochromatic wave 单色波
monochromator 单色仪,单色器,单色光镜
monoclinic 单斜晶的
monoclinic crystal 单斜晶体
monocrystal 单晶(体)
monolayer 单层
monomer 单体
monopole 单极子
monotropic liquid crystal 单变型液晶
Monte Carlo method 蒙特卡罗法
moon 月球,月亮
Morse alphabet 莫尔斯电码
Morse recorder 莫尔斯电报机
Moseley's law 莫塞莱定律
MOSFET (metal-oxide-semiconductor field effect transistor) 金属-氧化物-半导体场效应晶体管,MOS 场效管
Mössbauer effect 穆斯堡尔效应
most probable distribution 最概然分布
most probable speed 最概然速率,最可几速率
most probable value 最概然值,最可几值
MOS transistor (metal-oxide-semiconductor transistor) 金属-氧化物-半导体晶体管,MOS 晶体管
mother liquor 母液
mother solution 母液
motion 运动,移动

motional electromotive force 动生电动势
motor 马达,发动机,电动机
motor generator 电动发电机
mouse 鼠标
movable pulley 动滑轮
moving-coil instrument 动圈式仪表
moving-iron instrument 动铁式仪表
m. p.（melting point）熔点
MTF（modulation transfer function）调制传递函数
multiaccess system 多路存取系统
multichannel analyzer 多道分析器
multichannel spectrometer 多道分光计,多道谱仪
multifilamentary composite superconductor 多芯复合超导体
multilayer adsorption 多层吸附
multilayer dielectric film 多层介电膜,多层介质膜
multimedia 多媒体
multimeter 万用表,多用电表
multimode cavity 多模腔
multimode laser 多模激光器
multimode optical fibre 多模光纤
multiphase 多相
multiphoton 多光子
multiple beam 多光束,多光柱
multiple grating 复光栅
multiple scattering 多重散射
multiple star 〈天文〉聚星
multiplet 多重线,多重态
multiplication factor 增殖因数,倍增因数
multiplicity 多重性,多重数
multiplier 扩程器,倍增器;〈数〉乘数
multipole radiation 多极辐射
multipole transition 多极跃迁
multi-rotation 变异旋光
multistage rocket 多级火箭
multivibrator 多频振荡器
mu-meson μ子
Mumetal 高导磁合金
Muntz metal 熟铜,蒙次黄铜,蒙次合金
muon μ子
Murphy formula 墨菲公式
musical scale 音阶
mutarotation 变异旋光,变旋现象,旋光改变作用
mutual coherence 互相干性
mutual conductance 互导
mutual energy 互能
mutual inductance 互感系数
mutual induction 互感(应)
mutual inductor 互感器
muzzle velocity 腔口速度
mw（medium wave）中波
myopia 近视

Nn

N (newton) 牛顿(力的单位)
NAA (neutron activation analysis) 中子活化分析
nadir 〈天文〉天底
naked singularity 裸奇异性
NAND circuit 与非电路
nanometre (nm) 纳米,毫微米(长度单位)
nanosecond (ns) 纳秒(时间单位)
nanotechnology 纳米技术
Napierian logarithm 〈数〉自然对数,纳皮尔对数
narrow-band 窄带
narrow pulse 窄脉冲
natural abundance 自然丰度
natural coordinates 自然坐标
natural crystal plane 自然晶面
natural current 自然电流
natural damping 自然阻尼,固有阻尼
natural frequency 固有频率
natural light 自然光
natural logarithm 〈数〉自然对数
natural oscillation 固有振动,固有振荡
natural philosophy 自然哲学
natural radioactivity 天然放射性
natural unit 自然单位
natural vibration 自然振动,固有振动
natural wavelength 固有波长,自然波长
nautical mile 海里(长度单位)
Navier-Stokes equation 纳维-斯托克斯方程
NC (neutral current) 中性流
neap tide 低潮,弦潮,小潮
near axial ray 近轴光线
near field pattern 近场图样
near infrared 近红外
nearly axial ray 近轴光线
near point 近点
nearsightedness 近视
near ultraviolet 近紫外
nebula 〈天文〉星云
nebular hypothesis 〈天文〉星云假说
Néel temperature 奈尔温度
negative charge 负电荷,阴电荷
negative crystal 负晶体
negative deviation 负偏差
negative dislocation 负位错
negative energy state 负能态
negative feedback (NF) 负反馈
negative ion 负离子,阴离子
negative lens 负透镜
negative mass 负质量
negative meson 阴介子,负介子
negative plate 负极板,阴极板;底片
negative polarity 负极性
negative pole 负极,阴极
negative proton 阴质子,负质子
negative pulse 负脉冲
negative resistance diode 负阻二极管
negative sequence impedance 逆相序阻抗,负序阻抗
negaton 负子,电子
negatron 阴电子,负电子
negentropy 负熵
nematic crystal 向列型晶体
neon lamp 氖灯,霓虹灯
neper 奈培(衰减单位)

neptunium (Np) 〈化〉镎
neptunium series 镎系
Nernst effect 能斯特效应
Nernst theorem 能斯特定理
nesistor 双极场效应晶体管
net charge 净电荷
net force 净力
Neumann formula 诺埃曼(互感)公式
neutral 中性的,不带电的
neutral atom 中性原子
neutral current (NC) 中性流
neutral equilibrium 中性平衡,随遇平衡
neutrino 中微子
neutron 中子
neutron activation analysis (NAA) 中子活化分析
neutron bomb 中子弹
neutron cycle 中子循环
neutron death 中子俘获
neutron detector 中子探测器
neutron diffraction 中子衍射
neutron diffusion 中子扩散
neutron excess 中子过剩
neutron flux 中子通量
neutron generator 中子发生器
neutronics 中子(物理)学
neutron instrument 中子仪器
neutron inventory 中子总数
neutron irradiation 中子辐照
neutron leakage 中子泄漏
neutron mean free path 中子平均自由程
neutron multiplication 中子增殖
neutron number 中子数
neutron optics 中子光学
neutron reaction 中子反应
neutron scattering 中子散射
neutron source 中子源
neutron spectrograph 中子摄谱仪
neutron spectrometer 中子谱仪
neutron spectroscopy 中子谱学
neutron spectrum 中子能谱
neutron sponge 中子吸收剂
neutron star 〈天文〉中子星
neutron temperature 中子温度
new star 新星〈天文〉
newton (N) 牛顿(力的单位)
Newton first law 牛顿第一定律
Newtonian fluid 牛顿流体
Newtonian mechanics 牛顿力学,经典力学
Newtonian telescope 牛顿望远镜
Newton's disk 牛顿色盘
Newton second law 牛顿第二定律
Newton's formula 牛顿公式
Newton's law of cooling 牛顿冷却定律
Newton's law of gravitation 牛顿引力定律
Newton's laws of motion 牛顿运动三定律
Newton's rings 牛顿环;牛顿条纹
Newton third law 牛顿第三定律
NF (negative feedback) 负反馈
Nichrome 镍铬合金
nickel-cadmium cell 镍镉电池
nickel-iron battery 镍铁电池
nickel silver 德银,铜镍锌合金
Nicol prism 尼科耳棱镜
NiFe battery 镍铁电池
night vision 夜视
nit 尼特(光亮度单位)
nm (nanometre) 纳米,毫微米(长度单位)
NMR (nuclear magnetic resonance) 核磁共振
No (nobelium) 〈化〉锘
nocturnal radiation 红外辐射,夜间辐射
nodal point 节点
nodal slide 测节器,节点滑座
node 波节,节点,结点
Noether's theorem 诺特定理
no-hair theorem 无毛定理
noise 噪声,噪音,噪扰
noise abatement 噪声抑制,噪声降低
noise amplitude 噪声幅度
noise audiogram 噪声波形图,声

掩蔽听力图
noise background 本底噪声,噪声背景
noise diode 噪声二极管
noise dose 噪声剂量
noise dose meter 噪声剂量计
noise energy 噪声能量
noise environment 噪声环境
noise factor 噪声系数,噪声因数
noise figure 噪声系数,噪声因数
noise filter 噪声滤波器,静噪滤波器
noise function 噪声函数
noise immunity 抗扰度,抗扰性
noise insulation factor 隔音系数,隔音度
noise killer 噪声消除器
noise margin 噪声容限
noise meter 噪声计,噪声测试器
noise output 噪声输出
noise pollution 噪声污染
noise pulse 噪声脉冲
noise reduction 噪声降低
noise source 噪声源,噪扰源
noise suppression 噪声抑制
noise suppressor 噪声抑制器
noise trap 声阱,噪声陷波器
nominal deviation 标称偏差
nominal power 标称功率,额定功率
nominal value 额定值
nomogram 列线图,诺模图
nonadiabatic 非绝热的
non-central force 非有心力
noncoherent radiation 非相干辐射
non-conservative force 非保守力
noncrossing rule 不相交定则
noncrystal 非晶体
noncubic 非立方(系)的
nondestructive inspection 无损检测
nondimensionalization 无量纲化
nonequilibrium carrier 非平衡载流子
nonequilibrium state 非平衡态
nonequilibrium statistical mechanics 非平衡统计力学
nonequilibrium thermodynamics 非平衡热力学
non-Euclidean geometry 非欧几里得几何学
noninertial system 非惯性系
nonlinear effect 非线性效应
nonlinear optics 非线性光学
nonlinear response 非线性响应
nonlocalized system 非定域系
nonmetal 非金属(元素)
non-Newtonian fluid 非牛顿流体
nonpolar compound 〈化〉无极化合物
nonpolarized light 非偏振光
nonpolar molecule 无极分子
nonsingular matrix 〈数〉非奇异矩阵
non-spherical lens 非球面透镜
non-static process 非静态过程
nonstationary state 非定态
non-steady state 非稳态
nonstoichiometric compound 非化学计量化合物
nonthermal radiation 非热辐射
non-uniform motion 非匀速运动,变速运动
nonviscous flow 无黏流动
NOR gate 或非门
normal 〈数〉法线
normal acceleration 法向加速度,正交加速度
normal cell 标准电池
normal coordinate 简正坐标
normal dispersion 正常色散
normal distribution 正态分布
normal force 法向力,正交力
normal frequency 简正频率
normal incidence 正入射,垂直入射
normalization 归一化
normalizing factor 归一(化)因子
normal mode 简正模(式)
normal mode of vibration 简正振动

normal state 正常态,标准状态
normal stress 法向应力,正应力,正胁强
normal temperature and pressure (N. T. P.) 常温常压
normal vibration 简正振动
NOT circuit 非电路
note 律音,音符
NOT gate 非门
nova 〈天文〉新星
nova outburst 〈天文〉新星爆发
nox 诺克斯(照度单位)
Np (neptunium) 〈化〉镎
N pole 北极
NPR (nuclear paramagnetic resonance) 核顺磁共振
n-region n 区
ns (nanosecond) 纳秒(时间单位)
N. T. P. (normal temperature and pressure) 常温常压
n-type conductivity n 型电导率
n-type semiconductor n 型半导体
nuclear charge 核电荷
nuclear charge number 核电荷数
nuclear decay 核衰变
nuclear device 核装置
nuclear energy 核能,原子能
nuclear fission 核裂变,核分裂
nuclear force 核力
nuclear fuel (原子)核燃料
nuclear fusion 核聚变
nuclear heat 核热
nuclear isomerism 核同质异能性
nuclear magnetic resonance (NMR) 核磁共振
nuclear magnetism 核磁性
nuclear magneton 核磁子
nuclear mass 核质量
nuclear model 核模型
nuclear moment 核矩
nuclear paramagnetic resonance (NPR) 核顺磁共振
nuclear particle 核粒子,核质点
nuclear physics 核物理学,原子核物理学
nuclear power 核动力
nuclear power station 核电站
nuclear proliferation 核扩散
nuclear proton 核质子
nuclear radiation 核辐射
nuclear radius 核半径
nuclear reaction 核反应
nuclear reactor 核反应堆
nuclear resonance 核共振
nuclear shell model 核壳层模型
nuclear spectroscopy 核谱学
nuclear spin 核自旋
nuclear structure 核结构
nuclear war 核战争
nuclear warhead 核弹头
nuclear waste 核废料
nuclear weapon 核武器
nucleation 成核,晶核生成
nucleon 核子
nucleonics 核子学,原子核物理学
nucleonium 核子偶素
nucleon number 核子数
nucleor 裸核子,核子核心
nucleosynthesis 核合成
nucleus 原子核,核
nuclide 核素
null indicator 示零器
null method 衡消法,零位法
null vector 零(模)矢
numerical aperture 数值孔径
nutation 章动

Oo

object 物(体);靶,目标
object beam 物光束
object distance 物距
objective 物镜
objective grating 物端光栅
objective lens 物镜
objective lens adjustment 物镜调整
objective measurement 客观量度
objective prism 物端棱镜
object plane 物平面
object space 物(方)空间
oblique angle 斜角
oblique impact 斜碰,非对心碰撞
oblique incidence 斜入射
oblique lattice 斜点阵
oblique plane 斜面
oblique ray 斜光线,轴外光线
obliquity 倾角;倾斜
obliquity factor 倾斜因数,倾斜因子
observation 观察,观测
observation error 观测误差
observation port 观察孔
observatory 天文台,气象台,观测站
obstacle 障碍物
obstacle avoidance laser radar 激光防撞雷达
occlusion 吸留
occultation 〈天文〉掩,掩星;隐蔽,隐藏
octal notation 八进制计数法,八进位法
octane number 辛烷数,辛烷值
octave 倍频程;八度音阶
octave band 倍频带
octave filter 倍频程滤波器
octet 八位(位)组,八比特组
ocular 目镜
odd-even nucleus 奇偶核
odd-odd nucleus 奇奇核
odd parity 奇宇称,负宇称
Oe (oersted) 奥(斯特)(磁场强度单位)
off-axis aberration 轴外像差
off-axis angle 偏轴角
offset 偏置,偏移;补偿
offset angle 偏斜角
offset beam 漂移射束
offset cam 偏心凸轮
offset correction 偏差校正
offset crank mechanism 偏心曲柄机构
offset current 补偿电流
offset error 偏移误差
offset eyepiece 偏移目镜;补偿目镜
offsetting 偏置法
ohm(Ω) 欧(姆)(电阻单位)
ohmic contact 欧姆接触
ohmic heating 欧姆加热,电阻加热
ohmic loss 欧姆损耗,电阻损耗
ohmic resistance 欧姆电阻
ohmmeter 欧姆表,电阻表
Ohm's law 欧姆定律
oil capacitor 油浸电容器
oil condenser 油浸电容器
oil damping 油液减震
oil diffusion pump 油扩散泵
oil immersion lens 油浸透镜
oil immersion objective 油浸物镜
oil vapour 油雾

Olbers' paradox 奥伯斯佯谬,光度佯谬
omega-minus particle Ω负粒子
omegatron 回旋质谱仪
omnidirectional aerial 全向天线
opacity 不透明性,阻光性,不透明度
op-amp (operational amplifier) 运算放大器
open circuit 开路,断路
open cluster 〈天文〉疏散星团
open-hearth furnace 平炉
opening angle 开度角,孔径角
open system 开系,开放系统
open traverse 支导线,不闭合导线
open universe 开宇宙
opera glasses 小型双眼望远镜,观剧镜
operational amplifier (op-amp) 运算放大器
operator 算子,算符
ophthalmic optics 眼科光学,人眼光学
opposing e. m. f. 反电动势
opposite acting force 反向作用力
opposite parity 反宇称性
opposite phase 反相,逆相
opposite spins 反向自旋
opposition 〈天文〉冲
optical activity 旋光性;光学活性
optical aerial 光学天线
optical alignment 光学调准,光学准直
optical altimeter 光学测高仪
optical ammeter 光学电流计,光学安培计
optical amplification 光放大
optical amplitude modulator 光调幅器
optical analogue computer 光学模拟计算机
optical angle 光轴角
optical axis 光轴
optical beam flying 光束扫描
optical bench 光具座
optical biasing 光偏置
optical cable 光缆
optical centre 光(学中)心
optical collector 聚光器,聚光镜
optical communication 光通信
optical computer 光学计算机
optical constant 光学常数,光学恒量
optical crystal 光学晶体
optical cycle 光学周期
optical damage 光损伤
optical defect 光学缺陷,光疵
optical detector 光学探测器
optical device 光学器件,光学装置
optical diode 光二极管
optical disk 光盘,光具盘
optical distance 光程
optical distortion 光学畸变
optical double star 光学双星,视双星
optical element 光学元件
optical emission spectrograph 发光摄谱仪
optical fibre 光导纤维,光纤
optical fibre communication 光纤通信
optical flat 平晶,光学平面
optical Fourier transform 光学傅里叶变换
optical glass 光学玻璃
optical grating 光栅,光学栅格
optical grinding 光学研磨
optical guidance 光制导
optical heterodyne 光(学)外差
optical heterodyne radar 光频外差雷达
optical heterodyne receiver 光频外差接收器
optical holography 光学全息术
optical homer 光学(自动)跟踪头
optical homodyne 光(学)零差
optical image 光学图像
optical index 光学指数
optical information processing 光学信息处理

optical instrument 光学仪器
optical integration 光学集成
optical isomer 旋光异构体
optical length 光学长度,光程
optical lever 光杠杆
optically denser medium 光密介质
optical maser 激光器
optical medium 光学介质,光学媒质
optical memory 光学存储;光存储器
optical microscope 光学显微镜
optical mode 光学模
optical parallelism 光学平行性
optical path 光程
optical path difference 光程差
optical phonon 光频声子,光学声子
optical processing 光学处埋
optical pulse 光脉冲
optical pumping 光泵激,光抽运
optical pyrometer 光学高温计,光测高温计
optical radiation 光辐射
optical range finder 光学测距仪
optical reflector 反光镜
optical resolution 光学分辨率,光分辨能力
optical resonator 光学共振腔
optical rotary dispersion (ORD) 旋光色散
optical rotation 旋光性,旋光度
optical sensor 光学传感器,光敏感器
optical shutter 光闸,光学快门
optical system 光具组;光学系统
optical transfer function 光学传递函数,光学转移函数
optical transform 光学变换
optical transmittance 光透射率
optical trapping 光(学)陷俘
optic axis 光轴
optics 光学
optoacoustic effect 光声效应
optoelectronics 光电子学
optronics 光电子学
orbit 轨道
orbital 轨函数;轨道
orbital electron 轨道电子
orbital electron capture 轨道电子俘获
orbital quantum number 轨道量子数
orbital speed 轨道速度,第一宇宙速度
orbital velocity 轨道速度,第一宇宙速度
OR circuit 或电路
ORD (optical rotary dispersion) 旋光色散
order 级;有序;〈数〉阶
order-disorder transition 有序无序转变
order of diffraction 衍射级
order of interference 干涉级,干涉序
order of magnitude 数量级
order of spectrum 光谱级
order parameter 序参数
ordinary ray 寻常射线,寻常光线
ordinate 纵坐标
orientation 取向
origin 原点
orthogonal axis 正交(坐标)轴
orthogonal function 〈数〉正交函数
orthogonality 正交性,正交状态
orthogonalization 正交化
orthohydrogen 正氢
orthorhombic 正交的,斜方晶系的
orthoscopic eyepiece 无畸变目镜
orthoscopic image 无畸变图像
orthostigmat 广角镜头
oscillating crystal 振荡晶体;回摆晶体
oscillating current 振荡电流
oscillating dipole 振荡偶极子
oscillating universe 振荡宇宙
oscillation 振动,摆动,振荡
oscillator 振荡器;振子

oscillograph 示波器
oscilloscope 示波器
osmometer 渗压计
osmosis 渗透作用,渗透
osmotic pressure 渗透压
Ostwald ripening 奥斯特瓦尔德催热
Ostwald's dilution law 奥斯特瓦尔德稀释定律
Otto engine 奥托发动机
ounce（oz）盎司,英两(质量单位)
outgoing channel 出射道
outgoing particle 出射粒子
outgoing wave 出(射)波
out-of-focus image 离焦像
out of phase 异相,失相
output 输出,输出额;引出线
output coupling 输出耦合
output energy 输出能量
output impedance 输出阻抗
output meter 输出功率计
output power 输出功率
output pulse 输出脉冲
output quantity 输出量
output resistance 输出电阻
overdamping 过阻尼
overheating 过热
overlapping of orders 级的交叠
overload 过载
overpotential 超电势,超电位
overtone 泛音
overvoltage 过电压
overvoltage protection 过电压保护
overweight 过重,超重
oxide semiconductor 氧化物半导体
oz（ounce）盎司,英两(质量单位)
ozone concentration 臭氧浓度
ozone layer 臭氧层
ozone sphere 臭氧层
ozonolysis〈化〉臭氧分解
ozonometer 臭氧计
ozonosphere 臭氧层

Pp

Pa[1] (pascal) 帕(斯卡)(压强单位)
Pa[2] (protactinium) 〈化〉镤
packet of electrons 电子包,电子群
packet of energy 能包
packing density 组装密度,堆集密度,填充密度
packing fraction 敛集率
packing ring 垫圈,密封环,轴封环
paddlewheel satellite 带翼卫星,帆板卫星
pair creation 对产生
pair distribution function 对分布函数,偶分布函数
pair production 对产生
palaeomagnetism 古地磁学
panradiometer 全波段辐射计
pantograph 缩放仪,比例绘图器
pantometer 经纬测角仪
pantoscope 广角照相机,广角透镜
paper tape 纸带
parabola 〈数〉抛物线
parabolic reflector 抛物面反射镜
paraboloid 〈数〉抛物面
paraboloidal mirror 抛物面镜
paraboloidal reflector 抛物面反射器
parachor 等张比容
paracurve 抛物线
paradox 佯谬
parallactic angle 视差角,星位角
parallactic displacement 视差位移
parallactic ellipse 视差椭圆
parallactic error 视差误差
parallactic shift 视差移动
parallactoscopy 视差镜术
parallax 视差
parallax correction 视差修正
parallel axis theorem 平行轴定理
parallel beam 平行光束,平行射线束
parallel circuits 并联电路
parallelepiped 〈数〉平行六面体
parallel flow 平行流,层流
parallel force 平行力
parallelogram of forces 力的平行四边形
parallelogram of vectors 向量的平行四边形
parallelogram of velocities 速度的平行四边形
parallel processing 并行处理
parallel resonance 并联共振
parallel-series circuit 并-串联电路
parallel spins 平行自旋
paramagnetic moment 顺磁矩
paramagnetic resonance (PMR) 顺磁共振
paramagnetic susceptibility 顺磁磁化率
paramagnetism 顺磁性
paramagnon 顺磁波子,顺磁振子
parameter 参数,参量
parameter curve 参数曲线
parameter space 参量空间
parameter transformation 参数变换
parametric amplifier 参量放大器

parametric equation 参数方程
parametric excitation 参变激发，参量激励
parametric fluorescence 参变荧光
paraphase 倒相
paraphase amplifier 倒相放大器
parasite capacitance 寄生电容
parasitic capture 寄生俘获
parasitic effect 寄生效应
parasitic element 寄生元件，无源元件
parasitic feedback 寄生反馈
parasitic image 寄生像，杂像
parasitic loss 寄生损失，附加损失
parasitic oscillation 寄生振荡
parasitic reflector 寄生反射器，无源反射器
parasitic resonance 寄生共振，干扰共振
parasitics 寄生现象，寄生效应
parasitic scattering 寄生散射
parasitic suppressor 寄生振荡遏抑器
parastat 防静电器
paraxial 傍轴的，近轴的
paraxial approximation 傍轴近似
paraxial condition 傍轴条件
paraxial equation 傍轴方程
paraxial focus 傍轴焦点
paraxial imagery 傍轴成像
paraxial ray 傍轴光线
paraxial region 傍轴区，近轴区
parent 母体
parent nucleus 母核
parent structure 母结构
parity 宇称；奇偶性
parity bit 奇偶校验位
parity check 宇称检验；奇偶检验
parity conservation 宇称守恒
parity nonconservation 宇称不守恒
parity of wave function 波（动）函数的宇称（性）
parsec 秒差距（天体距离单位）
partial 泛音
partial derivative 〈数〉偏导数，偏微商
partial differential equation 〈数〉偏微分方程
partial eclipse 〈天文〉偏食
partial lunar eclipse 〈天文〉月偏食
partially coherent light 部分相干光
partial polarized light 部分偏振光
partial pressure 分压力，分压强
partial solar eclipse 〈天文〉日偏食
particle 质点；粒子
particle accelerator 粒子加速器
particle beam 粒子束
particle concentration 粒子浓度
particle diameter 颗粒直径，粒子直径
particle dynamics 质点动力学
particle formalism 微粒论
particle physics 粒子物理学
particle spectrum 粒子谱
particle velocity 质点速度
particulate matter 颗粒物质
partition 配分，划分；间壁
partition coefficient 配分系数
partition function 配分函数
parton 部分子
pascal (Pa) 帕（斯卡）（压强单位）
Pascal's law 帕斯卡定律
Paschen-Back effect 帕邢-巴克效应
Paschen series 帕邢系
passband 通（频）带，传输频带
passivation 钝化
passive device 被动器件，无源器件
path 路程，路径
path difference 程差
path integral 路径积分
path length 程长
path of integration 积分路线

path of ray 光路
Pauli exclusion principle 泡利不相容原理
PC (personal computer) 个人计算机,个人电脑
PCB (printed circuit board) 印刷电路板
PCM (pulse-code modulation) 脉冲编码调制
p.d. (potential difference) 势差,位差
peak 波峰;峰值,最大值
peak energy 峰能,峰值能量
peaker 峰化器
peak factor 峰值因数
peak height 峰(值)高(度)
peaking circuit 脉冲修尖电路
peak limiter 峰值限制器
peak load 峰值载荷
peak power 峰值功率
peak pulse amplitude 峰值脉冲幅度
peak response 峰值响应
peak strip 峰值带
peak value 峰值,最大值
pearlite 珠光体
peculiar star 特殊恒星
Peltier effect 佩尔捷效应
pendulum 摆
pendulum meter 摆式计数器
pendulum motion 摆运动
penetrating particle 穿透粒子,贯穿粒子
penetrating power 贯穿本领,穿透能力
penetrating progeny 次级穿透粒子,次级贯穿粒子
penetrating shower 贯穿簇射
penetration depth 穿透深度
penetration of current 电流穿透深度
penetration probability 穿透概率
penetrometer 透度计
pentode 五极管
penumbra 半影
perfect absorber 完全吸收体
perfect black body 完全黑体,绝对黑体
perfect conductor 理想导体
perfect crystal 完美晶体,无缺陷晶体
perfect diamagnetism 完全抗磁性
perfect differential 全微分
perfect elastic collision 完全弹性碰撞
perfect gas 完全气体,理想气体
perfect imaging 理想成像
perfect inelastic collision 完全非弹性碰撞
perfect lens 理想透镜
perfect radiator 完全辐射体
perfect set 完备集
perfect solution 理想溶液
performance factor 性能因数
periapsis 〈天文〉近(拱)点
periastron 〈天文〉近星点
pericynthion 〈天文〉近月点
perigee 〈天文〉近地点
perihelion 〈天文〉近日点
perilune 〈天文〉近月点
period 周期
periodicity 周期性
periodic law 周期律
periodic motion 周期运动
periodic table of elements 元素周期表
peripheral device 外设,外围设备
periscope 潜望镜
permalloy 坡莫合金,高导磁合金
permanent gas 永久气体
permanent magnet 永磁体
permanent magnetic field 永磁场
permanent magnetism 永磁性
permeability 渗透性,渗透率;磁导率
permeameter 磁导计
permissible overload 容许过载
permissible stress 容许应力
permitted transition 容许跃迁
permittivity 介电常量,介电常数,电容率

permutation 〈数〉排列
perpetual motion 永恒运动
perpetual motion machine 永动机
perpetual motion machine of the first kind 第一类永动机
perpetual motion machine of the second kind 第二类永动机
persistence of vision 视觉暂留
persistent current 持续电流
personal computer（PC）个人计算机，个人电脑
perturbation 摄动，微扰，微动
perturbation theory 微扰论
perversion （图像）左右倒置，横向倒置
PET（positron emission tomography）正电子发射层析术
pewter 白镴
Pfund series 普丰德系，普丰德级数
pH 〈化〉pH 值
phase 相，位相，相位
phase advance 相位超前
phase angle 相角
phase change 相变
phase changer 换相器，变相器
phase coherence 相位相干性
phase comparator 相位比较器
phase compensation 相位补偿
phase constant 相位常量
phase contrast 相衬，相位反差
phase-contrast microscope 相衬显微镜
phase coordinates 相坐标
phase corrector 相位校正器
phase curve 相位曲线
phase debunching 位相散乱
phase demodulator 相位解调器
phase diagram 相图
phase difference 相（位）差
phase discriminator 相鉴频器，相位鉴别器
phase displacement 相（位）移
phase distortion 相位失真，相畸变
phase equilibrium 相平衡
phase grating 相（位型）光栅，相栅
phase integral 相积分
phase inverter 倒相器，反相器
phase jump 相位跃变，相位跳变
phase lag 相位滞后
phase lead 相位超前
phase-locking 锁相
phase matching 相位匹配
phase meter 相位计
phase modulation 调相，相位调制
phaser 激声；移相器
phase resolution 相位分辨率
phase retardation 相位延迟
phase rule 相律
phase shift 相移
phase shifter 移相器
phases of the moon 月相
phase space 相空间
phase speed 相速（度）
phase synchronization 相位同步
phase transition 相变
phase velocity 相速（度）
phasing 定相
phasor 相矢量
phasotron 稳相加速器
phlogiston 燃素，热素
phlogiston theory 燃素学说
phon 方（响度单位）
phonometer 声强计
phonon 声子
phonon spectrum 声子谱
phosphor 磷光体
phosphor bronze 〈冶金〉磷青铜
phosphorescence 磷光
phot 辐透，厘米烛光（照度单位）
photino 光微子
photoacoustic effect 光声效应
photoacoustic spectroscopy 光声光谱学
photoactive 光敏的，光活的
photoactor 光电变换器件
photoadsorption 光致吸附

photo-beat 光拍，光频差拍
photocathode 光阴极，光电阴极
photocell 光电池；光电管；光电元件
photochemical reaction 光化反应
photochemistry 光化学
photochromic effect 光（赋）色效应
photochromic glass 光致变色玻璃
photochromism 光（赋）色性，光致变色，光致变色性
photochronograph 恒星中天摄影术，照相记时仪，活动物体照相机
photocomposition 照相排版
photocon 光（电）导元件，光导器件
photoconduction 光电导
photoconductive effect 光电导效应
photoconductive resistance 光敏电阻
photoconductivity 光电导性，光电导率
photoconductor 光电导体
photocurrent 光电流
photod 光电二极管
photodetachment 光致分离，光致脱离
photodetection 光（电）探测
photodetector 光电探测器
photodichroism 光致二向色性
photodiode 光电二极管，光敏二极管
photodisintegration 光致蜕变
photodissociation 光致离解，光解（作用）
photoeffect 光（电）效应
photoelastic effect 光弹效应
photoelasticity 光（测）弹性；光测弹性学
photoelectric cell 光电池，光电管，光电元件
photoelectric current 光电流
photoelectric effect 光电效应
photoelectric equation 光电效应方程
photoelectricity 光电；光电学；光电现象
photoelectric photometer 光电光度计
photoelectric threshold 光电阈
photoelectroluminescence 光控电致发光，光控场致发光
photoelectromotive force 光电动势
photoelectron 光电子
photoelectronics 光电子学
photoelectron spectroscopy 光电子能谱学，光电子谱学
photoelement 光电管，光生伏打电池
photoemission 光电子发射
photoengraving 照相制版
photoetching 光刻（蚀）
photoexcitation 光致激发
photofission 光致裂变
photoflash lamp 闪光灯
photofluorogram 荧光屏（图像）照片
photofluorography 荧光照相，荧光摄影术
photogalvanic effect 光（致）电流效应
photograph 照片，相片
photographic density 照相密度
photographic lens 照相镜头，照相透镜
photographic telescope 天体照相仪，照相望远镜
photography 照相术，摄影学
photogun 光电子枪
photohole 光穴
photoionization 光致电离，光化电离
photolithography 光刻，光刻法；平印照相制版
photology 光学
photoluminescence 光致发光，光激发光
photolysis 光解作用，光分解

photomagnetic effect 光磁效应
photometer 光度计
photometric brightness 光度学亮度
photometry 光度学,光度测定,测光
photomicrograph 显微照片
photomicrography 显微照相术
photomicrometer 测微光度计
photomixing 光混频
photomultiplier 光电倍增管,光电倍增器
photomultiplier tube 光电倍增管
photon 光子
photon bunching 光子群聚,光子集束
photon counting 光子计数
photon detector 光子探测器
photon effect 光子效应
photon energy 光子能量
photoneutron 光中子,光激中子
photon flux 光子通量
photonic crystal 光子晶体
photonics 光子学
photonuclear reaction 光(致)核反应
photopeak 光电峰
photoperiod 光周期
photophoresis 光致迁动,光致漂移,光泳
photopic vision 适亮视觉,白昼视觉
photoplastic effect 光塑效应,光范性效应
photoproton 光质子
photoradiogram 光电传真电报
photoreceptor 光感受体
photoresist 光致抗蚀剂;光阻材料
photoresistance 光敏电阻
photoresistor 光敏电阻(器)
photosensitivity 光敏性,感光性,感光灵敏度
photosensitization 光敏作用
photosphere 〈天文〉光球,光球层
photostar 光星,发光星体
photoswitch 光控继电器;光控开关
phototelegram 传真电报
phototelegraphy 传真电报术
phototelephone 传真电话机
phototransistor 光电晶体管,光敏晶体管,光敏三极管
phototube 光电管
photovaristor 光敏电阻(器),光电变阻器
photovoltaic cell 光电池,光生伏打电池
photovoltaic effect 光生伏打效应
physical balance 物理天平
physical chemistry 物理化学
physical optics 物理光学
physical pendulum 物理摆,复摆
physical quantity 物理量
physics 物理学
physisorption 物理吸附
pi(π) 圆周率
pie chart 饼形图
pi-electron π电子
piezocrystal 压电晶体
piezoelectric ceramics 压电陶瓷
piezoelectric crystal 压电晶体
piezoelectric effect 压电效应
piezoelectric oscillator 晶体振荡器,压电振荡器
piezoelectric quartz 压电石英
piezoelectrics 压电体
piezoelectric transducer 压电换能器
piezoelectric vibrator 压电振子
piezoid 压电石英
piezomagnetic effect 压磁效应
piezometer 压强计
piezo-optic effect 压光效应
piezoresistive effect 压致电阻效应,压阻效应
pig iron 生铁
pi-meson π介子
pinch effect 箍缩效应
pinhole camera 针孔照相机
pinhole imaging 针孔成像

pion π介子
Pirani gauge 皮拉尼真空规,皮拉尼真空计,皮拉尼压力计
piston 活塞
piston motion 活塞运动
piston rod 活塞杆
pitch 齿距;螺距;音调
Pitot tube 皮托管,空速管
pivot 枢轴,支点
pixel 像元,像素
planar array 平面阵列
planar defect 面缺陷
planar diode 平面二极管
planar factor 晶面因数
planar junction 平面结
planar technology 平面工艺
planar transistor 平面晶体管
Planck constant 普朗克常量
Planck formula 普朗克公式
Planck length 普朗克长度
Planck mass 普朗克质量
Planck's hypothesis 普朗克假说
Planck's radiation law 普朗克辐射定律
Planck units 普朗克单位
plane 平面;飞机
plane grating 平面光栅
plane hologram 平面全息图
plane mirror 平面镜
plane of incidence 入射(平)面
plane of polarization 偏振面
plane of symmetry 对称面
plane polarization 平面偏振
plane polarized light 平面偏振光
planet 〈天文〉行星
planetary motion 〈天文〉行星运动
planetoid 〈天文〉小行星
plane wave 平面波
plane wave Born approximation 平面波玻恩近似
planimeter 面积仪;求积仪
plano-concave lens 平凹透镜
plano-convex lens 平凸透镜
plasma 等离子体
plasma accelerator 等离子体加速器
plasma diffusion 等离子体扩散
plasma diode 等离子体二极管
plasma discharge 等离子体放电
plasma echo 等离子体回波
plasma etching 等离子体刻蚀
plasma focus 等离子体焦点
plasma frequency 等离子体频率
plasmaguide 等离子体波导
plasma gun 等离子体枪
plasma heating 等离子体加热
plasma laser 等离子体激光器
plasma linac 等离子体直线加速器
plasma oscillation 等离子体振荡
plasmapause 等离子体层顶
plasma physics 等离子体物理学
plasma sheath 等离子体鞘
plasmasphere 等离子体层
plasmatron 等离子体管,等离子体发生器
plasmoid 等离子体团
plasmon 等离体子
plastic crystal 塑性晶体
plastic deformation 塑性形变,范性形变
plasticity 塑性,范性;塑性力学
plate 板;板极;极板
plateau 坪区,台阶
plateau diffusion 坪区扩散
plateau region 坪区
plate rectifier 板形整流器
pleochroism 多向色性
pliotron 功率三极管
plumb line 铅垂线,重垂线
plunger 柱塞
plutonium (Pu)〈化〉钚
PMR (paramagnetic resonance) 顺磁共振
p-n junction p-n 结
p-n junction diode p-n 结二极管
p-n junction laser p-n 结激光器
Po (polonium)〈化〉钋
Poincaré group 庞加莱群
Poincaré sphere 庞加莱球
point 尖端;点

point charge 点电荷
point-contact junction 点接触结
point-contact transistor 点接触式晶体管
point defect 点缺陷
point discharge 尖端放电
pointer 指针;指示器
point group 点群
point lattice 点晶格,点阵
point particle 点粒子
point symmetry 点对称
poise[1] 平衡;砝码
poise[2] 泊(黏滞度单位)
Poiseuille law 泊肃叶定律
Poisson bracket 泊松括号
Poisson distribution 泊松分布
Poisson's equation 泊松方程
Poisson's ratio 泊松比
polar bond 〈化〉极性键
polar coordinates 〈数〉极坐标
polar crystal 极性晶体
polar distance 极距
polarimeter 旋光计,偏振计
polariscope 偏(振)光镜
polarity 极性
polarizability 极化性,极化率,极化度
polarization 极化,极化度;偏振
polarization charge 极化电荷
polarization current 极化电流
polarization curve 极化曲线
polarization direction 偏振方向;极化方向
polarization effect 极化效应;偏振效应
polarization ellipse 偏振椭圆
polarization of light 光的偏振
polarizer 起偏器,起偏(振)镜,偏振元件
polarizing angle 起偏角
polarizing plate 起偏振片
polar molecule 〈化〉极性分子,有极分子
polarograph 极谱记录仪
polaroid 偏振片
polaron 极化子
polar vector 极矢(量)
polaxis 极化轴
pole 磁极;电极;极;〈数〉极点
pole distance 极距
pole effect 电极效应
pole piece 极靴,磁极片
pole strength 电极强度;磁极强度;极强
politure 抛光
pollution 污染
poloidal magnetic field 角向磁场
polonium (Po) 〈化〉钋
polychromatic light 多色光
polychromatic radiation 多色辐射
polycrystal 多晶(体)
polydomain 多畴
polyelectrode 多电极
polyelectron 多电子
polygon 〈数〉多角形,多边形
polygon of forces 力多边形
polyhedral angle 多面角
polyhedron 多面体
polymer 〈化〉聚合物,聚合体
polymer physics 高分子物理学
polymer semiconductor 聚合物半导体
polynomial 〈数〉多项式;多项的
polysilicon 多晶硅
polytropic exponent 多方指数
polytropic process 多方过程
poor conductor 不良导体
population inversion 粒子数反转,粒子数布居反转,布居反转
position vector 位置矢量,位矢
positive charge 正电荷,阳电荷
positive crystal 正晶体
positive feedback 正反馈
positive ion 正离子,阳离子
positive plate 正极板
positive polarity 正极性
positive pole 正极,阳极
positron 正电子,阳电子
positron emission tomography (PET) 正电子发射层析术
positronium 电子偶素,正电子素

post 接线柱,端子
post-deflection acceleration 偏转后加速
potential 势,位
potential barrier 势垒,位垒
potential difference (p.d.) 势差,位差
potential divider 分压器,电位器,电势分配器
potential drop 电势降,势降,位降
potential energy 势能
potential energy surface 势能面
potential field 势场,位场
potential function 势函数,位函数
potential hill 势垒,位垒
potential scattering 势散射
potential theory 位势论
potential trough 势阱,位阱
potentiometer 电位差计,电势差计,电位器
pound 磅(质量单位)
poundal 磅达(力的单位)
powder metallurgy 粉末冶金学
power 功率;焦度;〈数〉幂
power density spectrum 功率密度谱
power dissipation 功率耗散,功率消耗
power factor 功率因数
power function 〈数〉功效函数,幂函数
power input 功率输入
power installation 动力装置,动力站
power klystron 功率速调管
power line 电力传输线,输电线
power loss 功率损耗
power magnetron 大功率磁控管
power meter 功率表,功率计
power output 功率输出
power pack 动力单元;电源组
power plant 发电厂,发电站
power reactor 动力反应堆
power rectifier 功率整流器
power semiconductor device 功率半导体器件
power source 电源
power spectrum 功率谱
power station 发电厂,发电站
power supply 电源
power transmission 电力传输
power tube 功率管
Poynting vector 坡印亭矢量
preaccelerator 前加速器
preamplification 前置放大
preamplifier 前置放大器
precession 进动,旋进
precessional motion 进动
precession of the equinoxes 〈天文〉分点岁差
precipitation 淀积,脱溶
precise instrument 精密仪器
precision 精密度
preheating 预热
preimage 原像
preinjector 预注入器
preionization 预电离
preon 前子
preprocessing 预处理
presbyopia 远视眼,老花眼
present 现在
pressure 压强,压力
pressure altimeter 气压测高计
pressure coefficient 压力系数,压强系数
pressure compensator 压力补偿器
pressure curve 压力曲线
pressure difference 压力差,压强差
pressure diffusion 压力扩散,压强扩散
pressure drag 压差阻力
pressure drop 压降,压致损失
pressure effect 压力效应
pressure equation 压力方程
pressure field (大气)压力场
pressure gauge 压力计,压强计
pressure meter 压力计,压力表
pressure shift 压致频移,压致

移动
pressure stress 压胁强,压应力
pressure switch 压力开关,压强开关
pressure-temperature phase diagram 压(强)温(度)相图
pressure transducer 压力传感器
pressure transmitter 压力传送器,压强传送器
pressurized cabin 加压舱;密封舱
pressurized-water reactor (PWR) 压水(核)反应堆
Prevost's theory of heat exchange 普雷伏热交换理论
primary aberration 初级像差
primary cell 原电池
primary colour 原色
primary image 原像
primary mirror 主镜
primary optical axis (晶体)主光轴
primary winding 原绕组
prime focus 主焦点
prime meridian 本初子午圈,本初子午线
primitive cell 原胞
principal axis 主轴
principal axis of inertia 惯量主轴
principal focus 主焦点
principal maximum 主极大
principal moment of inertia 主转动惯量
principal plane 主面
principal point 主点
principal quantum number 主量子数
principal ray 主光线
principal refractive index 主折射率
principal series 主线系
principal stress 主胁强,主应力
principal vector 主矢(量),主向量
principal velocity 主速度
principal wave 主波,基波
principle of constancy of light velocity 光速不变原理
principle of entropy increase 熵增加原理
principle of equivalence 等效原理
principle of general relativity 广义相对性原理
principle of indeterminacy 测不准原理,不确定原理
principle of relativity 相对性原理
principle of special relativity 狭义相对性原理
principle of superposition 叠加原理
principle of virtual displacement 虚位移原理
principle of virtual work 虚功原理
printed circuit 印刷电路
printed circuit board (PCB) 印刷电路板
printer 打印机
prism 棱镜
prismatic spectrum 棱镜光谱
prism spectrograph 棱镜摄谱仪
probability 概率,几率
probability current 概率流
probability density 概率密度,几率密度
probability function 概率函数,几率函数
probability integral 概率积分,几率积分
probable error 概然误差
probable velocity 概然速度,可几速度
probe technique 探针技术
product of inertia 惯量积
product phase 子相
program 程序
programmable read-only memory (PROM) 可编程只读存储器
progressive wave 前进波
projectile 抛体;射弹

projectile motion 抛体运动
projection 投影;投掷
projection objective 投影物镜
projection operator 投影算符
projective transformation 射影变换
projector 投影仪;投光灯
proliferation 增殖
PROM (**programmable read-only memory**) 可编程只读存储器
prompt neutron 瞬发中子
propagation 传播
propagation constant 传播常量
propagation of error 误差传递
propagation vector 传播矢量
propagation velocity 传播速度
propagator 传播函数
propeller 推进器
proper energy 原能
proper function 本征函数,特征函数
proper length 静长度,固有长度
proper mass 固有质量,静止质量
proper motion 〈天文〉固有运动,自行
proper time 固有时
proper value 本征值,特征值
proportion 〈数〉比例
proportional counter 正比计数器
proportional limit 比例极限
protactinium (**Pa**) 〈化〉镤
protogalaxy 〈天文〉原星系
proton 质子
proton decay 质子衰变
protonium 质子偶素
proton number 质子数,原子序数
proton resonance 质子共振
proton spin resonance 质子自旋共振
proton synchrotron 质子同步加速器
protostar 〈天文〉原恒星
pseudopotential 赝势
pseudoscalar 赝标量
pseudotensor 赝张量
pseudovector 赝矢量,伪矢量
psi particle ψ 粒子
psychrometer 干湿球湿度计,干湿计
Ptolemaic system 托勒密(宇宙)体系
p-type semiconductor p 型半导体
Pu (**plutonium**) 〈化〉钚
pulley 滑轮
pulley blocks 滑轮组
pulsar 〈天文〉脉冲星
pulsatance 角频率,圆频率
pulsation 脉动
pulse-code modulation (**PCM**) 脉冲编码调制
pulsed laser 脉冲激光器
pulse duration 脉冲宽度,脉冲持续时间
pulse jet 脉动式喷气发动机
pulse modulation 脉冲调制
pulser 脉冲发生器
pulse radar 脉冲雷达
pulse reflection 脉冲反射
pulse spacing 脉冲间隔
pulse spectrum 脉冲谱
pulse strength 脉冲强度
pulse stretching 脉冲展宽
pulse transmission 脉冲传输
pulse triggering 脉冲触发
pulse width 脉冲宽度,脉冲持续时间
pump 泵;抽运,抽吸,汲取
pupil 光瞳,瞳孔
pupil function (光)瞳函数
pure rolling 纯滚动,完全滚动
pure state 纯态
pure strain 单纯胁变,单纯应变
purification 提纯
purity 纯度
push-pull 推挽式的
push-pull amplifier 推挽式放大器
PWR (**pressurized-water reactor**) 压水(核)反应堆
pyramid 角锥,棱锥体

pyroelectric effect 热电效应
pyroelectricity 热电，热电学
pyrometer 高温计
pyrometric cone 示温熔锥，测温三角锥，泽格测温锥
pyrometry 高温测量学
pyrophoricity 自燃

Qq

QCD (quantum chromodynamics) 量子色动力学
QED (quantum electrodynamics) 量子电动力学
Q factor (quality factor) Q 因数，品质因数
QFD (quantum flavour dynamics) 量子味动力学
QGD (quantum gravitational dynamics) 量子引力动力学
Q-modulation 调 Q，Q 调制
q-number q 数
QSG (quasi-stellar galaxy) 类星星系
QSO (quasi-stellar object) 类星体
Q-spoiling Q 突变
QSS (quasi-stellar radio source) 类星射电源
Q-switch Q 开关，光量开关
qt (quart) 夸脱（容量单位）
quadratic equation 二次方程
quadrode 四极管
quadrupler 四倍频器
quadrupole 四极子
quadrupole moment 四极矩
quality factor (Q factor) Q 因数，品质因数
quality of sound 音品，音色，音质
quantization 量子化
quantum 量子
quantum chaos 量子混沌
quantum chemistry 量子化学
quantum chromodynamics (QCD) 量子色动力学
quantum corral 量子围栏
quantum correction 量子修正
quantum cosmology 量子宇宙学
quantum defect 量子数亏损
quantum dot 量子点
quantum effect 量子效应
quantum electrodynamics (QED) 量子电动力学
quantum electronics 量子电子学
quantum field theory 量子场论
quantum flavour dynamics (QFD) 量子味动力学
quantum gravitational dynamics (QGD) 量子引力动力学
quantum gravity 量子引力
quantum Hall effect 量子霍尔效应
quantum mechanics 量子力学
quantum number 量子数
quantum optics 量子光学
quantum physics 量子物理学
quantum simulation 量子模拟
quantum size effect 量子尺寸效应
quantum state 量子态
quantum statistics 量子统计法
quantum theory 量子理论
quantum theory of radiation 辐射的量子理论
quantum transition 量子跃迁
quantum well 量子阱
quantum yield 量子产额
quark 夸克
quark confinement 夸克禁闭
quart (qt) 夸脱（容量单位）
quarter-wave plate 1/4 波片
quartz 石英
quartz clock 石英钟
quartz oscillator 石英振荡器
quasar 恒星状球体，类星体

quasicrystal 准晶
quasi-equilibrium 准平衡
quasi-Fermi level 准费米能级
quasi-fission 准裂变
quasilinear equation 准线性方程
quasi-molecule 准分子
quasi-particle 准粒子
quasi-static process 准静态过程
quasi-stellar galaxy (QSG) 类星星系
quasi-stellar object (QSO) 类星体
quasi-stellar radio source (QSS) 类星射电源
quenching 淬火
Q-value Q值,优值

Rr

Ra (radium) 〈化〉镭
racemate 外消旋混合物
racemic mixture 外消旋混合物
rad[1] 拉德(辐射吸收剂量单位)
rad[2] (radian) 弧度
radar (radio detection and ranging) 雷达(无线电探测和测距)
radial acceleration 径向加速度
radial distribution function 径向分布函数
radial momentum 径向动量
radial quantum number 径量子数
radial structure function 径向结构函数
radial sweep 径向扫描
radial velocity 径向速度
radian (rad) 弧度
radiance 辐射亮度,辐射率
radiant emittance 辐射发射度
radiant energy 辐射能
radiant exitance 辐射出射度
radiant exposure 辐射量
radiant flux 辐射通量
radiant flux density 辐射通量密度
radiant heat 辐射热
radiant intensity 辐射强度
radiant power 辐射功率
radiant transmittance 辐射透射率
radiation 辐射,放射
radiation belt 辐射带,范艾伦带
radiation coefficient 辐射系数
radiation damage 辐射损伤
radiation dose 辐射剂量
radiation hazard 辐射危害
radiation impedance 辐射阻抗
radiation length 辐射长度
radiationless decay 无辐射衰减
radiationless transition 无辐射跃迁
radiation modification 辐射改性
radiation pattern 辐射(方向)图
radiation pressure 辐射压(强)
radiation protection 辐射防护
radiation resistance 辐射电阻
radiation safety 辐射安全
radiation shield 辐射屏蔽
radiation source 辐射源
radiation temperature 辐射温度
radiative absorption 辐射吸收
radiative capture 辐射俘获
radiative collision 辐射碰撞
radiativity 辐射性,放射性
radiator 辐射体
radical 〈数〉根;根式;〈化〉根
radio 无线电,收音机
radioactive 放射性的,有辐射的
radioactive age 放射性年龄
radioactive dating 放射性鉴年法
radioactive decay 放射衰变
radioactive equilibrium 放射性平衡
radioactive fallout 放射性沉降物,放射性散落物,辐射微尘,原子尘
radioactive isotope 放射性同位素
radioactive nuclide 放射性核素
radioactive series 放射系
radioactive source 放射源
radioactive tracer 放射性示踪

物,放射性指示剂
radioactive-tracer method 放射示踪法
radioactive waste 放射性废物
radioactivity 放射性,放射性活度
radio astronomy 射电天文学
radiobiology 放射生物学
radio broadcasting 无线电广播
radiocarbon dating 放射性碳定年,放射性碳计时
radiochemistry 放射化学
radio communication 无线电通信,射电通信
radio detection and ranging (radar) 雷达(无线电探测和测距)
radioelectronics 无线电电子学
radio-frequency (RF) 射频
radio-frequency heating 射频加热
radio-frequency oscillator 射频振荡器
radio-frequency spectrum 射频(波)谱
radio galaxy 射电星系
radiogenic 放射产生的
radiography 射线照相术
radio interferometer 无线电干涉仪
radioisotope 放射性同位素
radiolocation 无线电定位
radiology 辐射学,放射学
radioluminescence 辐射(致)发光
radiolysis 射解作用,辐解作用
radionuclide 放射性核素
radio-opacity 辐射不透明度
radio receiver 无线电接收机,收音机
radiosonde 无线电探空仪
radio source 无线电源,射电源
radio star 射电星
radio telescope 射电望远镜,无线电天文望远镜
radiotherapy 放射治疗
radio transmission 无线电传递
radio window 无线电窗
radium (Ra) 〈化〉镭
radius of curvature 曲率半径
radius of gyration 回旋半径
radius vector 〈数〉径矢,径向量
rainbow 虹霓
RAM (random-access memory) 〈计算机〉随机存取存储器
r.a.m. (relative atomic mass) 〈化〉相对原子质量
Raman effect 拉曼效应
Raman laser 拉曼激光器
Raman scattering 拉曼散射
Raman spectrum 拉曼(光)谱
ramjet 喷气引擎
ramp 斜面,斜坡
Ramsden eyepiece 拉姆斯登目镜
random-access memory (RAM) 〈计算机〉随机存取存储器
random motion 无规运动,随机运动
random walk 无规行走,随机游动
range 量程,射程
Rankine cycle 兰金循环
Rankine temperature scale 兰氏温标
Raoult's law 拉乌耳定律
rapid cooling 急冷
rapidity 快度
rare-earth metal 稀土金属
rare-earth permanent magnet 稀土永磁体
rarefaction 稀疏,稀化
rarefaction wave 稀疏波
rarefied plasma 稀薄等离(子)体
rare metal 稀有金属
raster 光栅,网格
rated voltage 额定电压
rate meter 定率计
rationalized unit 合理单位
rational number 〈数〉有理数
raw data 原始数据
ray 光线,射线
ray diagram 射线图
Rayleigh criterion 瑞利判据

Rayleigh-Jeans formula 瑞利-金斯公式
Rayleigh number 瑞利数
Rayleigh scattering 瑞利散射
ray tracing 光线追迹
r.d.(relative density)相对密度
reactance 电抗
reaction 反作用;反动力;反应
reaction energy 反应能
reaction force 反作用力
reaction principle 反作用原理
reaction propulsion 反作用推进
reaction Q-value 反应Q值
reactive current 无功电流
reactive power 无功功率
reactor 反应堆
read-only memory(ROM)只读存储器
real gas 真实气体
real image 实像
Réaumur temperature scale 列氏温标
recalescence 〈冶金〉复辉,再辉
reciprocal lattice 倒易点格,倒易点阵
reciprocal lattice point 倒格点
reciprocal lattice vector 倒格矢
reciprocal space 倒易空间
recoil 反冲,反弹
recoil electron 反冲电子
recombination 复合
reconstruction 重构
rectangular-well potential 矩阱势
rectification 整流;纠正
rectifier 整流器
rectilinear motion 直线运动
rectilinear propagation 直线传播
red dwarf 〈天文〉红矮星
red giant 〈天文〉红巨星
redshift 〈天文〉红移
reduced mass 约化质量,折合质量
reed switch 簧片开关
reference beam 参考光束
reference frame 参考系,参照系
reference system 参考系,参照系
reflectance 反射比,反射系数
reflected ray 反射线
reflected wave 反射波
reflecting telescope 反射望远镜
reflection 反射
reflection angle 反射角
reflection coefficient 反射系数
reflection grating 反射光栅
reflection law 反射定律
reflectivity 反射率,反射系数;反射性
reflector 反射镜,反射器,反射望远镜
refracted ray 折射线
refracting telescope 折射式望远镜
refraction 折射
refraction angle 折射角
refraction coefficient 折射系数
refraction law 折射定律
refractive constant 折射常数,折光常数
refractive index 折射率,折光率
refractivity 折射本领,折射率;折射性
refractometer 折射计
refractor 折射器;折射式望远镜
refractory 耐熔质,耐火材料;难熔的,耐火的
refrigerant 制冷剂,冷冻剂
refrigeration 制冷,冷冻
refrigerator 制冷器,制冷机,电冰箱
regelation 再凝,复冰
regression analysis 回归分析
regular reflection 规则反射,镜(面)反射
rejection of data 数据舍弃
relative acceleration 相对加速度
relative aperture 相对孔径,相对口径
relative atomic mass(r.a.m.)〈化〉相对原子质量
relative density(r.d.)相对密度
relative error 相对误差
relative humidity 相对湿度

relative molecular mass 分子量
relative motion 相对运动
relative permeability 相对渗透率;相对磁导率
relative permittivity 相对电容率
relative velocity 相对速度
relativistic 相对论性的
relativistic effect 相对论(性)效应
relativistic field equation 相对论性场方程
relativistic mass 相对论(性)质量
relativistic mechanics 相对论(性)力学
relativistic particle 相对论性粒子
relativistic physics 相对论(性)物理学
relativistic quantum mechanics 相对论(性)量子力学
relativity 相对性;相对论
relativity of simultaneity 同时性的相对性
relativity principle 相对性原理
relativity theory 相对论
relaxation 弛豫
relaxation function 弛豫函数
relaxation oscillator 张弛振荡器
relaxation time 弛豫时间
relay 中继;继电器
reliability 可靠性
relief telescope 体视望远镜
reluctance 磁阻
reluctivity 磁阻率
rem (roentgen equivalent man) 雷姆(人体伦琴当量)
remanence 剩磁,顽磁
remanent polarization 剩余极化强度
remote control 遥控
remote sensing 遥感,遥测
renewable energy source 再生能源,可更新能源
renormalization 重正化
renormalization group 重正化群
replica grating 复制光栅
reproducibility 可重复性,可复现性,再现性
reptation effect 蠕动效应
repulsive force 推斥力,斥力
residual aberration 剩余像差
residual image 余像,余留像,残留图像
resistance 电阻;阻力
resistance box 电阻箱
resistance pyrometer 电阻高温计
resistance thermometer 电阻温度计
resistivity 电阻率
resistor 电阻(器)
resolution 分辨率;分解
resolution of force 力的分解
resolution of velocity 速度分解
resolving power 分辨本领,分辨率
resolving time 分辨时间
resonance 谐振,共振;共鸣;共振态
resonance absorption 共振吸收
resonance box 共振箱
resonance energy 共振能量
resonance fluorescence 共振荧光
resonance frequency 共振频率
resonance integral 共振积分
resonance level 共振级
resonance peak 共振峰
resonance spectrum 共振谱
resonance state 共振态
resonance width 共振宽度
resonant cavity 共振腔,谐振腔
resonant circuit 共振电路
resonant mode 共振(波)模,共振模式
resonator 共振器,共鸣器
response curve 响应曲线
response function 响应函数
rest energy 静能
restitution coefficient 恢复系数
rest mass 静质量
restoring force 回复力
restricted theory of relativity 狭

义相对论
resultant 合力;合量
resultant couple 合力偶
resultant force 合力
resultant velocity 合速度
retardation 减速,延迟
retardation plate (相位)延迟板,推迟板,减速板
retentivity 剩磁,顽磁性
retina 视网膜
retrograde motion 〈天文〉逆行
retrorocket 制动火箭,减速火箭
reverberant sound 混响声
reverberation 混响;反射
reverberation time 混响时间
reverberator 反射器,反射镜
reverberatory furnace 反射炉
reversed image 反像
reverse feedback 负反馈
reverse osmosis 逆渗透
reverse reaction 逆反应
reverse thrust 反推力,逆推力
reversibility of optical path 光路可逆性
reversible cycle 可逆循环
reversible engine 可逆机
reversible process 可逆过程
reversing prism 反像棱镜,倒像棱镜
revolution 循环;绕转
Reynolds number 雷诺数
RF (**radio-frequency**) 射频
rhe 流值(流度单位)
rheology 流变学
rheonomic constraint 非定常约束
rheopexy 触变性
rheostat 变阻器,可变电阻器
rheotaxial growth 液相外延生长
rheotron 电子感应加速器,电磁感应加速器
rhombohedral lattice 菱形点阵,菱面体格,三方点格
Richardson-Dushman equation 理查孙-德西曼方程
Richardson equation 理查孙方程
Richter magnitude 里氏震级,里克特震级
right-hand circular polarization 右(旋)圆偏振
right-handed crystal 右旋晶体
right-handed current 右手流
right-handed screw rule 右手螺旋定则
right-hand rule 右手定则
rigid body 刚体
rigidity 刚度系数,刚度
rigidity modulus 刚性模量
rigid rod 刚性杆
rigid structure 刚性结构
ring circuit 环形电路,环路
ring main 环路,环形电路
ring mode 环形模
ripple 波纹;波动;波纹度
rishon 初子
RMS (**root mean square**) 方均根
RMS current 方均根电流
RMSE (**root mean square error**) 方均根误差
RMS speed 方均根速率
RMS value 方均根值
RMS voltage 方均根电压
robotics 机器人学
Rochelle salt 罗谢尔盐,酒石酸钾晶体
Rochon prism 罗雄棱镜
rock crystal 无色水晶
rocket 火箭
roentgen 伦琴(辐射剂量单位)
roentgen equivalent man (**rem**) 雷姆(人体伦琴当量)
rolling friction 滚动摩擦
rolling motion 滚动
ROM (**read-only memory**) 只读存储器
root mean square (**RMS**) 方均根
root mean square error (**RMSE**) 方均根误差
Rose's metal 罗斯合金,铅铋锡易熔合金
rot 旋度
rotary converter 旋转变流机
rotary pump 旋转泵

rotating mirror 旋(转)镜
rotation 转动;自转;旋度;旋光
rotational energy 转动能
rotational flow 旋流
rotational transform 旋转变换
rotation around a fixed point 定点转动
rotation axis 转动轴
rotation number 转数
roton 旋子
rotor 转子,回转轴,转动体
Rowland ring 罗兰环
ruby laser 红宝石激光器
rule 定则,定律
runaway electron 脱逸电子
Russell-Saunders coupling L-S 耦合,罗素-桑德斯耦合
Rutherford backscattering 卢瑟福背散射,卢瑟福后向散射
Rutherford-Bohr atom model 卢瑟福-玻尔原子模型
Rutherford experiment 卢瑟福(散射)实验
Rutherford scattering 卢瑟福散射
Rydberg atom 里德伯原子
Rydberg constant 里德伯常量,里德伯常数
Rydberg molecule 里德伯分子
Rydberg series 里德伯(谱线)系

Ss

S (siemens) 西门子(电导单位)
s (second) 秒(时间单位)
saddle point 鞍点
safelight 安全灯
safety clutch 安全离合器
safety condition 安全条件
safety factor 安全因数,安全因子
safety glass 安全玻璃,不碎玻璃
safety lamp 安全灯
safety valve 安全阀
sagittal focal line 弧矢焦线
sample function 样本函数
sampler 取样器,采样器
sample space 样本空间,抽样空间
sample value 样本值
sample variance 样本方差
sampling 取样
satellite 卫星
satellite line 伴线
satellite peak 伴峰
satellite structure 伴线结构
satellite wave 伴波
saturated gain 饱和增益
saturated polarization 饱和极化强度
saturated solution 饱和溶液
saturated vapour 饱和蒸汽
saturation 饱和;饱和值,饱和度
saturation current 饱和电流
saturation curve 饱和曲线
saturation intensity 饱和强度
saturation region 饱和区
saturation resistance 饱和电阻
saturation spectroscopy 饱和光谱学
saturation value 饱和值
saturator 饱和器
sawtooth generator 锯齿波发生器
sawtooth sweep 锯齿波扫描
sawtooth waveform 锯齿波形
scalar 标量,无向量
scalar electron 标量电子
scalar field 标量场
scalar flux 标量通量
scalar interaction 标量相互作用
scalar matrix 标量矩阵
scalar meson 标量介子
scalar meson theory 标量介子理论
scalar multiplication 标乘法
scalar pair theory 标量偶理论
scalar potential 标势
scalar product 内积,点积,标(量)积,数量积
scalar quantity 标量,无向量
scale 标度;比例尺;刻度尺;音阶;秤
scale eyepiece 带标目镜
scalene 不规则三角形,不等边三角形
scale parameter 标度参数
scale plate 刻度板
scaler 定标器;换算装置
scale transformation 标度变换
scale value 标度值,刻度值
scaling circuit 定标电路,计数电路
scaling law 标度律,标度无关律
scaling variable 标度无关变量
scanned imagery 扫描成像
scanned laser 扫描激光器
scanner 扫描器,扫描仪

scanning 扫描
scanning angle 扫描角
scanning aperture 扫描孔径
scanning beam 扫描束
scanning camera 扫描照相机
scanning electron microscope (SEM) 扫描电子显微镜
scanning interferometer 扫描干涉仪
scanning lens 扫描透镜
scanning monochromator 扫描单色仪
scanning prism 扫描棱镜
scanning tunnelling microscope (STM) 扫描隧穿显微镜
scattered light 散射光
scattered reflection 漫反射
scatterer 散射体,散射器
scattering 散射
scattering angle 散射角
scattering constant 散射常数,散射恒量
scattering cross-section 散射截面
scattering function 散射函数
scattering kernel 散射核
scattering length 散射长度
scattering loss 散射损失
scattering matrix (S-matrix) 散射矩阵,S矩阵
scattering nucleus 散射核
scattering of electromagnetic radiation 电磁辐射散射
scattering particles 散射粒子
scattering pattern 散射图样
scattering power 散射本领
scatterometer 散射仪
scatter trap 散射阱
schlieren photography 纹影摄影术
Schmidt camera 施密特照相机
Schmidt corrector (plate) 施密特校正板
Schottky barrier 肖特基势垒
Schottky defect 肖特基缺陷
Schottky effect 肖特基效应
Schottky noise 肖特基噪声,散粒噪声
Schrödinger equation 薛定谔方程
Schuler pendulum 舒勒摆
Schwarzschild radius 施瓦氏半径
Schwarzschild solution 施瓦氏解
Schwinger effect 施温格效应
scintillation 闪烁
scintillation camera 闪烁照相机
scintillation counter 闪烁计数器
scintillation crystal 闪烁晶体
scintillation spectrometer 闪烁谱仪
scintillator 闪烁器;闪烁体
scintillometer 闪烁计数器
scintilloscope 闪烁镜
scintilogger 闪烁测井计数器
scintiphotogram 闪烁照相图
scintiphotography 闪烁照相术
scintiscanner 闪烁扫描器
scintiscanning 闪烁扫描
scission point 断点
sclerometer 回跳硬度计
scleronomic constraint 定常约束
scleroscope 回跳硬度计
scleroscope hardness 回跳硬度
scotopic vision 适暗视觉,微光视觉
screen 屏;屏幕,荧光屏
screen grid 屏栅极
screen image 屏像
screw 螺旋,螺钉,螺杆
screw thread 螺纹
seaquake 海震,海啸
search coil 探察线圈
Searle conduction apparatus 瑟尔热导仪
second(s) 秒(时间单位)
secondary aberration 次级像差
secondary cell 蓄电池
secondary colour 合成色,副色
secondary effect 二次效应,次发效应
secondary electron 次级电子
secondary emission 次级发射
secondary ion mass spectroscopy

次级离子质谱学
secondary isotope 二级同位素
secondary maximum 次极大
secondary mirror 副镜
secondary motion 次级运动
secondary optical axis 副光轴
secondary partical 次级粒子
secondary photoelectric effect 次级光电效应
secondary radiation 次级辐射
secondary rainbow 霓,副虹
secondary resistance 次级电阻
secondary resonance 次级共振
secondary standard 副标准
secondary stress 次级胁强,次级应力
secondary wave 次波,S波
secondary wavelet 次级子波
secondary winding 次级绕组,副绕组
second Brillouin zone 第二布里渊区
second cosmic velocity 第二宇宙速度
second harmonic 二次谐波
second law of thermodynamics 热力学第二定律
second-order phase transition 二级相变
secular equation 久期方程,特征方程
secular equilibrium 长期平衡
sedimentation 淀积
Seebeck effect 泽贝克效应
Seger cone 泽格测温锥
seism 地震
seismic wave 地震波
seismograph 地震仪,测震仪
seismology 地震学
selection rule 选择定则
selective amplifier 选频放大器
selective excitation 选择激发
selective reflection 选择反射
selectron 标量电子
selenium cell 硒光电池
selenology 月球学
self-consistency 自洽性
self-consistent field 自洽场
self-consistent solution 自洽解
self-energy 自能
self-excited oscillation 自激振荡,自激振动
self-inductance 自感,自感系数
self-induction 自感(应)
self-organization 自组织
SEM (scanning electron microscope) 扫描电子显微镜
semibreadth 半宽度
semiclassical approach 半经典方法
semiclassical approximation 半经典近似
semiconductivity 半导性
semiconductor 半导体
semiconductor diode 半导体二极管
semiconductor element 半导体元件
semiconductor junction 半导体结
semiconductor laser 半导体激光器
semiconductor physics 半导体物理学
semicrystalline state 半晶态
semiempirical 半经验的
semipermeable membrane 半透膜
semiquantitative 半定量的
semireflector 半反射器
semistable state 亚稳态
sensitivity 灵敏度
sensitometer 感光计
sensor 传感器,敏感元件
separating prism 分像棱镜
separation energy 分离能
separatrix 分界线
sequence 时序,序列
series 〈数〉级数;串联
series circuits 串联电路
series element 串联元件
series resonance 串联共振
series winding 串激绕组;串激绕法

series-wound machine 串激电机
set 装置,仪器;集合
set theory 集合论
sextant 六分仪
shadow 影,阴影
shadow effect 阴影效应
shadow photometer 比影光度计
shadow region 阴影区
shadow scattering 核影散射
shaft 轴,杆状物
shallow junction 浅结
shallow level 浅能级
shape-memory alloy 形状记忆合金
shearing force 剪力
shearing interferometer 剪切干涉仪
shearing strain 剪应变,切应变
shear modulus 剪切模量
shell electron 壳层电子
shell model 壳(层)模型
sherardizing 渗锌粉,锌粉热镀
SHF (super high frequency) 超高频
shielded nuclide 屏蔽核素
shield grid 屏栅
shielding 屏蔽
shielding effect 屏蔽效应
SHM (simple harmonic motion) 简谐运动
Shockley energy level 肖克利能级
shock tube 击波管
shock wave 冲击波
short circuit 短路
short-range force 短程力
short-sightedness 近视
shower 簇射
shunt 分路;分流;分流器
shunt winding 并激绕组,并激绕法
Si (silicon) 〈化〉硅
sideband 边频带,旁频带
sidereal day 恒星日
sidereal period 恒星周期
siemens (S) 西门子(电导单位)
sievert (Sv) 西弗特(剂量当量单位)
sigma particle σ粒子
signal 信号,讯号
signal-to-noise ratio 信噪比
sign convention 符号规约
significant figures 有效数字
silicon (Si) 〈化〉硅
silicon chip 硅芯片,集成电路
silicon controlled rectifier 可控硅整流器
silicon diode 硅二极管
silicon dioxide 二氧化硅
silicon photocell 硅光电池
silicon photodetector 硅光电探测器
silicon photodiode 硅光电二极管
simple harmonic motion (SHM) 简谐运动
simple harmonic wave 简谐波
simple pendulum 单摆
simplex circuit 单工电路
simulation 仿真,模拟
sine curve 正弦曲线
sine rule 〈数〉正弦定律
sine wave 正弦波
single crystal 单晶(体)
single domain 单(磁)畴
single exposure 单次曝光
single mode 单模
single-mode fibre 单模光纤
single-pass gain 单程增益
single-sideband modulation 单边带调制
single-slit diffraction 单缝衍射
singlet state 单态
singularity 奇异性;奇点
singular perturbation 奇异扰动
sintering 烧结,热压结
sinusoid 正弦曲线
sinusoidal oscillator 正弦振荡器
sinusoidal wave 正弦波
siphon 虹吸;虹吸管
SI units (Système International d'Unités) 国际单位制
size effect 尺寸效应

skew symmetry 斜对称,反对称
skin depth 趋肤深度
skin effect 趋肤效应
skip distance 越距,越程
sky wave 天空电波
slepton 标量轻子
sliding friction 滑动摩擦
slip band 滑移带
slip of dislocation 位错滑移
slit 狭缝,裂缝
slow neutron 慢中子
slug 斯(勒格)(质量单位);铁芯
small-angle scattering 小角散射
S-matrix (scattering matrix) 散射矩阵,S矩阵
S-matrix theory S矩阵理论
smectic phase 层状相,近晶相,脂状相
smelting 熔炼
smoke 烟
smoothing capacitor 平滑电容器,滤波电容器
Snell's law 斯涅耳定律
Snell's law of refraction 斯涅耳折射定律
socket 插座,插口
sodium lamp 钠灯
sodium-vapour lamp 钠汽灯
soft iron 软铁
soft magnetic material 软磁材料
soft mode 软模
soft radiation 软辐射
software 软件
soft water 软水
sol 溶胶
solar cell 太阳能电池
solar constant 太阳常数,太阳恒量
solar day 〈天文〉太阳日
solar eclipse 〈天文〉日食
solar energy 太阳能
solar flare 〈天文〉太阳耀斑
solar magnetism 太阳磁;太阳磁学
solar panel 太阳电池板
solar parallax 太阳视差
solar physics 太阳物理学
solar prominence 〈天文〉日珥
solar radiation 太阳辐射
solar spectrum 太阳光谱
solar system 太阳系
solar telescope 太阳望远镜
solar tide 太阳潮
solar wind 太阳风
solar year 太阳年,回归年
solder 软钎料
solder drop junction 焊滴结
solenoid 螺线管
solid 固体
solid angle 立体角,多面角
solidification 凝固
solidifying point 凝固点
solid phase epitaxy 固相外延
solid solution 固溶体
solid-state circuit 固态电路
solid-state detector 固态探测器
solid-state physics 固态物理,固体物理
solid-state plasma 固态等离子体
soliton 孤(立)子
solstices 〈天文〉二至点
solubility 可溶性;溶度
solubility coefficient 溶解系数
solubility curve 溶线
solubility exponent 溶度指数
solubility limit 溶度限
solubility product 〈化〉溶度积
solute 溶解物,溶质
solution 溶液;溶解
solvent 溶剂
solvus 溶线,固溶线
Sommerfeld elliptic orbit 索末菲椭圆轨道
sonar (sound navigation and ranging) 声呐(声波导航与测距)
sonde 探头;探针;探空仪
sonic barrier 声障
sonic boom 声震
sonic depth finder 回音测深仪
sonics 声能学
sonic speed 声速

sonoluminescence 声致发光
sonometer 弦音计
sorption 吸附(作用)
sorption pump 吸附泵
sound 声(音)
sound absorption 吸声,隔声
sound barrier 音障
sound concentrator 聚声器
sound-intensity level 声强级
sound level meter 声级计
sound navigation and ranging (sonar) 声呐(声波导航与测距)
sound pressure 声压
sound pulse 声脉冲
sound radar 声波雷达
sound ray 声线
sound source 声源
sound spectrum 声谱
sound wave 声波
source 源;源点
space 空间;间隔
space charge 空间电荷
space charge region 空间电荷区
space correlation 空间关联
spacecraft 太空飞行器
space curve 空间曲线
space inversion 空间反演
space lattice 空间格点
space probe 空间探测器
space reflection 空间反射
spacetime 时空
spacetime continuum 时空连续统
spallation 散裂
spark chamber 火花室
spark counter 火花计数器
spark discharge 火花放电
spark erosion 火花电蚀
spark gap 火花隙
spark plug 电花插头,火花塞
spark quenching 火花淬灭
spark spectrum 火花光谱
spattering 溅射
speaker 扬声器
special relativity 狭义相对论
special theory of relativity 狭义相对论
specific activity 比活度,比放射性
specific binding energy 比结合能
specific charge 荷质比,比荷
specific gravity 比重
specific heat 比热
specific heat anomaly 比热反常
specific heat at constant pressure 定压比热
specific heat at constant volume 定体(积)比热
specific heat capacity 比热容
specific humidity 比湿
specific ionization 比电离度,电离比值
specific phase 特定相
specific resistance 比阻,电阻率,电阻系数
specific rotation 比旋,旋光率
specific surface 比(反应)面积
specific volume 比体积,比容
specific weight 比重
spectral analysis 光谱分析,频谱分析
spectral band (光)谱带
spectral characteristic 光谱特性
spectral class 谱型,谱类
spectral decomposition 频谱分解
spectral density 频谱密度
spectral function 谱函数
spectral intensity 光谱强度
spectral line (光)谱线
spectral order 光谱级
spectral purity 光谱纯度
spectral series (光)谱线系
spectral type 光谱型
spectrogram 光谱图
spectrograph 摄谱仪
spectrography 摄谱学
spectrometer 分光计,光谱仪
spectrophotometer 分光光度计
spectroscope 分光镜
spectroscopic binary 分光双星
spectroscopy 光谱学,光谱术,波谱学
spectrum 谱;波谱;光谱

spectrum of first order 一级光谱
spectrum selectivity 光谱选择性
specular reflection 镜(面)反射
speculum 制镜合金
speed 速率,速度
speed of light 光速
speed of sound 声速
speedometer 速率计
speed reducer 减速器
speed-time graph 速率-时间图
sphere 球体;〈天文〉天球
spherical aberration 球(面像)差
spherical harmonics 〈数〉球谐函数
spherical lens 球面透镜
spherical mirror 球面镜
spherical mirror resonator 球面镜共振腔
spherical nucleus 球形核
spherical polar coordinates 球极坐标
spherical projection 球面投影
spherical wave 球面波
spherometer 球径计
spiegel(eisen)〈冶金〉镜铁
spike 尖峰(信号);激光尖峰,脉冲激光
spike pulse 尖峰脉冲
spin 自旋
spin angular momentum 自旋角动量
spin degeneracy 自旋简并性
spin density 自旋密度
spin effect 自旋效应
spin energy 自旋能
spin glass 自旋玻璃
spinodal curve 旋节线,亚稳界线
spinodal decomposition 旋节线分离,旋节线分解,失稳分解
spinor 旋量
spin-orbit coupling 自旋轨道耦合
spin-orbit splitting 自旋轨道分裂
spin quantum number 自旋量子数
spin resonance 自旋共振
spin tensor 自旋张量
spin wave 自旋波
spin wave resonance 自旋波共振
spiral 螺(旋)线
spiral galaxy 螺旋星系,旋涡星系
split of energy level 能级分裂
split stream 分流
spontaneous emission 自发发射
spontaneous fission 自发裂变,自发(核)分裂
spontaneous magnetization 自发磁化;自发磁化强度
spontaneous transition 自发跃迁
spring 弹簧
spring balance 弹簧秤
spring tide 大潮,朔望潮
spurion 乱真子
spurious image 伪像,杂光像
sputtering 溅射
square root 〈数〉平方根
square wave 方形波
squark 标量夸克
sr (steradian) 球面度(立体角单位),立体弧度
stability 稳定性
stability criterion 稳定性判据,稳定性准则
stabilized current supply 稳流电源
stabilized voltage supply 稳压电源
stable equilibrium 稳定平衡
stable particle 稳定粒子
stacking fault 堆垛层错
stainless steel 不锈钢
standard atmospheric pressure 标准大气压
standard cell 标准电池
standard clock 标准钟
standard deviation 标准偏差
standard electrode 标准电极
standard electrode potential 标准电极电位
standard error 标准误差
standard form 标准型
standard model 标准模型

standard state 标准态
standard temperature and pressure (s.t.p.) 标准温度和压力
standing wave 立波,驻波
star 星,恒星
star atlas 〈天文〉星图,天图
star cloud 〈天文〉星云
star cluster 〈天文〉星团
Stark effect 斯塔克效应
state 态
state function 态函数
state of matter 物态
state parameter (物)态参量
state space 态空间
state variable 态变量
state vector 态矢量
static electricity 静电;静电学
static friction 静摩擦
static register 静态寄存器
statics 静力学
static stress 静胁强,静应力
stationary ensemble 定态系综,稳定系综
stationary equilibrium 稳定平衡
stationary field 定态场,恒定场
stationary orbit 稳定轨道,静止轨道,同步轨道
stationary particle 静粒子,驻粒子
stationary phase 稳相
stationary satellite 同步卫星
stationary Schrödinger equation 定态薛定谔方程
stationary state 定态,稳态
stationary wave 定态波,驻波
statistical average 统计平均,统计平均值
statistical equilibrium 统计平衡
statistical error 统计误差
statistical mechanics 统计力学
statistical physics 统计物理学
statistical regularity 统计规律性
statistical weight 统计权重
statistics 统计学,统计法
stator 定子;定片
steady constraint 定常约束
steady current 恒定电流,稳恒电流
steady flow 定常流(动),稳流
steady state 定常态,稳恒态
steady-state theory of the universe 定态宇宙理论
steam calorimeter 蒸汽量热器
steam distillation 蒸汽蒸馏
steam engine 蒸汽机
steam point 汽点
steam turbine 蒸汽(涡)轮机
steel 钢
Stefan-Boltzmann law 斯特藩-玻尔兹曼定律
Stefan's law 斯特藩定律
stellarator 仿星器
stellar evolution 恒星演化
stellar interferometer 恒星干涉仪
stellar magnitude 星等
stellar parallax 恒星视差
stellar wind 恒星风
step attenuator 步进衰减器
step-down transformer 降压变压器
step motor 步进电动机,步进马达
stepped attenuator 阶式减光板
stepped lens 阶梯透镜,分步透镜
stepped pulse 阶梯脉冲
stepped winding 阶梯绕组
stepping switch 步进开关
stepping tube 步进制电子管
step potentiometer 分级电位计
step pulse 阶梯脉冲
step regulator 步进调节器
step tube 步进制电子管
step-up transformer 升压变压器
steradian (sr) 球面度(立体角单位),立体弧度
stere 立方米(体积单位)
stereometer 体积计
stereophonic sound 立体声
stereoptics 立体光学,立体摄影光学

stereoscope 体视镜
stereoscopic film 立体电影
stereotelescope 体视望远镜
Stern-Gerlach experiment 施特恩-格拉赫实验
stick force 黏附力
sticking coefficient 黏附系数
stiffness 倔强性;劲度,刚度
stiffness coefficient 刚度系数,劲度系数
stiffness factor 倔强因数
stiffness modulus 劲度模量
stigmator 消像散器
stilbmeter 亮度计
stimulated absorption 受激吸收
stimulated emission 受激发射,感生发射
stimulated radiation 受激辐射,感生辐射
stimulated Raman scattering 受激拉曼散射
Stirling cycle 斯特林循环
Stirling engine 斯特林热气机
STM (scanning tunnelling microscope) 扫描隧穿显微镜
stochastic error 随机误差
stochastic process 随机过程
Stokes' law 斯托克斯定律
Stokes' parameters 斯托克斯参数
Stokes' theorem 斯托克斯定理
stop 光阑;停止
stopping distance 制动距离;阻止距离
stopping power 阻止本领
storage battery 蓄电池(组)
storage cell 蓄电池
storage ring 存储环
s.t.p. (standard temperature and pressure) 标准温度和压力
straightening 校直,调直
strain 应变,胁变
strain gauge 应变规,应变计
strain hardening 胁变硬化,应变硬化
strange attractor 奇怪吸引子
strangeness 奇异性
strange particle 奇异粒子
stratified medium 分层介质
straton 层子
stratosphere 同温层,平流层
stray capacitance 杂散电容
stray capacity 杂散电容
stray current 杂散电流
stray radiation 杂散辐射
streamline 流线;流线型
streamline flow 流线流,稳定流动
stress 应力,胁强
stress birefringence 应力双折射
stress hardening 应力硬化
stress tensor 应力张量,胁强张量
string 弦
string model 弦模型
stripped plasma 全剥等离子体,完全(电离的)等离子体
stroboscope 频闪仪,频闪观测仪
strobotach 频闪测速计
stroke 冲程
strong coupling 强耦合
strong interaction 强相互作用
strong turbulence 强湍动
structural phase transition 结构相变
structural relaxation 结构弛豫
structure factor 结构因子
structure function 结构函数
subatomic particle 亚原子粒子
subcritical 次临界的,亚临界的
subjective brightness 主观亮度
sublimate 升华物质
sublimation 升华
submillimetre wave 亚毫米波
subshell 支壳层
subsonic speed 亚声速
substate 亚级,亚能级,亚态
subtractive process 减色法
sun 日,太阳
sunspot 〈天文〉太阳黑子
superallowed transition 超容许跃迁
supercluster 〈天文〉超星系团

superconducting 超导的
superconducting film 超导膜
superconducting magnet 超导磁体,超导磁铁
superconducting magnetic levitation 超导磁浮
superconductivity 超导(电)性
superconductor 超导体
supercooling phenomenon 过冷现象
supercritical 超临界的
superdense star 〈天文〉超密(恒)星
superfluid 超流体
superfluidity 超流(动)性,超流态
supergalaxy 〈天文〉超星系
supergiant 〈天文〉超巨星
supergravity 超引力
superheavy element 超重元素
superheavy nucleus 超重核
superheterodyne receiver 超外差接收机
super high frequency (SHF) 超高频
superlattice 超点阵,超晶格,超格点
supernova 〈天文〉超新星
superparamagnetism 超顺磁性
superplasticity 超塑性
superposition principle 叠加原理
superpower laser 超功率激光器
superpressure 超大气压,超压
superproton 超质子,超高能质子
superradiance 超辐射
superradiation 超辐射
superrefraction 超折射
supersaturated solution 过饱和溶液
supersaturation 〈化〉过饱和
supersaturation vapour pressure 过饱和蒸汽压
supersonic 超声波的
supersonic generator 超声波发生器
supersonic speed 超声速,超音速
supersonic wave 超声波
superspace 超空间
superstring theory 超弦理论
supersymmetry 超对称(性)
supertransuranics 超重元素
superweak interaction 超弱相互作用
suppressor grid 抑制栅极
surd 〈数〉无理数,不尽根
surface active agent 表面活性剂
surface current 表面流;面电流
surface current density 面电流密度
surface depletion layer 表面耗尽层
surface energy 表面能
surface entropy 表面熵
surface hardening 表面硬化,表面淬火
surface integral (表)面积分
surface physics 表面物理学
surface state 表面态
surface stress 表面应力
surface structure 表面结构
surface tension 表面张力
surface vibration 表面振动
surface wave 表面波
surfactant 表面活性剂
surge current 浪涌电流,冲击电流
surge impedance 波阻抗,电涌阻抗
surge protection 过电压保护
susceptance 电纳
susceptibility 磁化率,极化率
suspension 悬置;悬浮
suspension bridge 悬桥
Sv (sievert) 西弗特(剂量当量单位)
sweep coil 扫描线圈
switch 开关,电键
switching time 开关时间
symmetry 对称(性)
synchrocyclotron 同步回旋加速器
synchronous 同步的

synchronous motor 同步电动机
synchronous orbit 同步轨道
synchronous rotation 同步旋转
synchronous satellite 同步卫星
synchrophasotron 同步稳相加速器
synchrotron 同步加速器
synchrotron radiation 同步(加速器)辐射
syngony 晶系
synodic month 〈天文〉朔望月,阴历月
synodic period 〈天文〉会合周期
synthetic aperture 综合孔径,合成孔径
systematic error 系统误差
Système International d'Unités (**SI units**) 国际单位制
systems software 〈计算机〉系统软件
syzygy 〈天文〉朔望

Tt

T (tesla) 特斯拉(磁感应强度单位)
tachometer 转速计
tachyon 快子(超光速粒子)
tandem accelerator 串列式加速器
tangent 〈数〉正切;切线
tangent galvanometer 正切电流计
tangential acceleration 切向加速度
tangential stress 切向应力,切向胁强
tangential velocity 切向速度
target 靶;目标
target chamber 靶室
target current 靶电流
target isotope 靶同位素
target nucleus 靶核
target particle 靶粒子
target strength 目标强度
target theory 靶理论,靶学说
tau lepton τ轻子
tau particle τ粒子
Taylor series 〈数〉泰勒级数
tektite 熔融石,玻璃陨石
telecentric 焦阑的,远心的
telecentric beam 远心光束
telecentric stop 焦阑,远心光阑
telecentric system 焦阑系统,远心系统
telecommunication 电信,长途通讯,长途电信
telecommunication satellite 通信卫星
telecontrol 遥控
telefax 电传
telegauge 遥测仪,测远计,测距仪
telegram 电报
telegraph 电报;电报机
telelens 远摄镜头,长焦镜头
telemechanics 遥控力学,遥控机械学
telemeter 测距仪,遥测仪
telemicroscope 望远显微镜
telephone 电话
telephoto 传真电报;远摄照相
telephoto lens 远摄镜头,长焦镜头
telephotometer 远距光度计,遥测光度计
telescope 望远镜
telescopic sight 望远瞄准器
telescopy 望远镜学;望远镜制造学
telescreen 电视银幕,荧光屏
teleseism 远震
teleset 电视(机)
telespectroscope 远距分光镜
telestar 电视卫星
telestereoscope 体视望远镜
teleswitch 遥控键,遥控开关
teletalking 有声电影
telethermometer 遥测温度计
teletorque 自动同步机,交流自整角机
teletron 显像管
teletyper 电传打字机,打字电报机
television (TV) 电视(机)
television camera 电视摄像机
television tube 电视显像管
televisor 电视(机)

telewriter 打字电报机
tellurometer 微波测距仪
temperature 温度
temperature coefficient 温度系数
temperature compensation 温度补偿
temperature compensator 温度补偿器
temperature conductivity 导热性,热导率
temperature correction 温度校正
temperature difference 温差
temperature dispersion 温度弥散
temperature drift 温度漂移
temperature drop 温度降差
temperature field 温度场
temperature gradient 温度梯度
temperature radiation 热辐射
temperature scale 温(度)标
temperature variation 温度变化
tempering 回火,退火
temporal coherence 时间相干性
temporary magnetism 暂(时)磁性
tensile elasticity 伸长弹性
tensile figure 拉伸指数,张力数
tensile force 张力
tensile modulus 拉伸模量
tensile strain 拉伸应变,张胁变,张应变
tensile strength 抗张强度,抗拉强度
tensile stress 张胁强,张应力,拉应力
tensility 张力,伸力,伸长性
tensimeter 气体压力计
tensiometer 张力计
tension 张力,拉力
tension dynamometer 张力测力计
tension spring 张丝,拉伸弹簧
tensodiffusion 张力扩散
tensometer 张力计
tensor 张量
tensor analysis 张量分析
tensor density 张量密度
tensor function 张量函数
tensor operator 张量算子
tensor transformation 张量变换
tephigram 温熵图
terminal 端子,线接头,极柱
terminal ballistics 终点弹道学
terminal condition 终端条件
terminal impedance 终端阻抗
terminal speed 终极速度,终极速率
terminal velocity 终极速度,终极速率
terminal voltage 端电压
terminator 〈天文〉明暗界限,晨昏线
terrestrial magnetism 地磁;地磁学
terrestrial radiation 地面辐射,地球辐射
terrestrial refraction 地面折射
terrestrial telescope 地面望远镜
tesla (T) 特斯拉(磁感应强度单位)
Tesla coil 特斯拉线圈
Tesla transformer 特斯拉变压器
test particle 试探粒子
tetragonal system 四角晶系,四方晶系
tetrahedron 四面体
tetrode 四极管
texture 织构;构造
Th (thorium) 〈化〉钍
theodolite 经纬仪
theorem 定理
theoretical error 理论误差
theoretical mechanics 理论力学
theoretical physics 理论物理学
theoretical strength 理论强度
theory 理论,学说
therm 克卡,小卡(热量单位)
thermal activation 热激活
thermal aftereffect 热后效
thermal agitation 热骚动,热扰动
thermal anchoring 热汇
thermal capacity 热容量
thermal conductivity 导热性;热

导率
thermal conductor 热导体
thermal diffusion 热扩散
thermal efficiency 热效率
thermal energy 热能
thermal equilibrium 热平衡
thermal equilibrium condition 热平衡条件
thermal equivalent 热当量
thermal equivalent of work 热功当量
thermal excitation 热激发,热激励
thermal expansion 热膨胀
thermal expansion coefficient 热膨胀系数
thermal farad 热法拉
thermal fatigue 热疲劳
thermal feedback 热反馈
thermal field 热场,温度场
thermal fluctuation 热涨落
thermal flux 热通量
thermal fragmentation 热裂(作用)
thermal hysteresis 热滞后
thermal image converter 热变像管
thermal imaging 热成像
thermal insulation 热绝缘,绝热
thermal insulator 热绝缘体,绝热器
thermalization 热能化
thermal lens 热透镜
thermal lensing effect 热透镜效应
thermal light 热光
thermal load 热负荷
thermal motion 热运动
thermal neutron 热中子
thermal radiation 热辐射
thermal radiator 热辐射体
thermal reactor 热反应堆
thermal resistance 热阻
thermal resistivity 热阻率,热阻系数
thermal resistor 热敏电阻
thermal spike 温度峰值,热峰
thermal stability 热稳定性
thermal state 热态
thermal stratification 热力层结
thermal stress 热胁强,热应力
thermal switch 热开关
thermal transmission 传热,热传递
thermal unit 热量单位
thermal value 热值
thermal valve 热阀
thermal velocity 热速度
thermal vibration 热振动
thermion 热离子,热电子
thermionic amplifier 热离子放大器
thermionic cathode 热离子阴极,热阴极
thermionic conduction 热离子传导
thermionic converter 热离子转换器
thermionic current 热离子电流
thermionic diode 热离子二极管
thermionic emission 热电子发射,热离子发射
thermionic rectifier 热离子整流器
thermionics 热离子学
thermionic tube 热离子管
thermionic valve 热离子管,热阴极电子管
thermistor 热敏电阻
thermite 铝热剂,灼热剂
thermoammeter 热电偶安培计
thermoanalysis 热学分析
thermoanelasticity 热滞弹性
thermobarometer 温度气压表
thermochemistry 热化学
thermocouple 热电偶,温差电偶
thermocouple gauge 热电偶真空计,温差电偶真空规
thermocouple thermometer 热电偶温度计
thermodiffusion 热扩散
thermodynamic equilibrium 热力

学平衡
thermodynamic factor 热力学因子
thermodynamics 热力学
thermodynamic scale 热力学温标
thermodynamic temperature 热力学温度
thermoelectric effect 热电效应，温差电效应
thermoelectricity 热电，温差电
thermoelectron 热电子
thermograph 温度记录器
thermography 温度记录术
thermoisopleth 等温线
thermojunction 热电偶接头
thermology 热学
thermoluminescence 热致发光
thermoluminescent dating 热致发光测定年代
thermomagnetic effect 磁场热流效应
thermometer 温度计
thermometric property 测温性质
thermometric scale 温标
thermometry 计温学，温度测定法
thermonuclear fusion 热核聚变
thermonuclear plasma 热核等离子体
thermonuclear reaction 热核反应
thermonuclear reactor 热核反应堆
thermonuclear weapon 热核武器
thermopile 温差电堆，热电堆
thermostat 恒温器，恒温箱
theta vacuum θ 真空
Thevenin theorem 戴维南定理
thin film interference 薄膜干涉
thin film optics 薄膜光学
thin lens 薄透镜
third cosmic velocity 第三宇宙速度
third law of thermodynamics 热力学第三定律
third order aberration 三级像差
thixotropy 触变性，摇溶现象
Thomson effect 汤姆逊效应
Thomson scattering 汤姆逊散射
thorium (Th)〈化〉钍
thorium series 钍系
three-body problem 三体问题
three-dimensional 三维的，立体的
three-phase alternating current 三相交变电流
threshold 阈，界限
threshold condition 阈值条件
threshold energy 阈能
threshold frequency 阈频(率)
threshold of feeling 痛阈
threshold of pain 痛阈
threshold power 阈值功率
throttle valve 节流阀
through hole 通孔
thrust 推力，冲度
thunderstorm 雷暴
thyratron 闸流管
thyristor 晶闸管，闸流晶体管
tidal energy 潮汐能
tide 潮汐
tight-binding approximation 紧束缚近似
timbre 音品，音色，音质
time 时间
time base 时基
time dila(ta)tion 时间延缓，时间膨胀，时间扩展
time-lapse photography 延时摄影
timelike 类时
timelike event 类时事件
timelike interval 类时间隔
timelike vector 类时矢量
timer 计时器，定时器
time response 时间响应
time reversal 时间反演，时间反转
time scale 时标
time scale factor 时标因子
time series 时间序列
time sharing 分时
time sweep 时间扫描

time vector 时间矢量
time zone 时区
timing circuit 计时电路,定时电路
timing pulse generator 定时脉冲发生器
T-invariance T不变性
Tokamak 托卡马克(受控热核反应装置)
tolerance 容限;公差
tolerance deviation 容许偏差
tomography 层析成像,层析X射线照相术
ton 吨(质量单位)
tone 音调
topology 拓扑学
toric lens 环面透镜,复曲面透镜
torque 转矩,扭矩
torr 托(真空度单位)
Torricellian vacuum 托里拆利真空
torsion 扭转;转矩;〈数〉挠率
torsional mode 扭转模
torsional rigidity 抗扭劲度,扭转刚度
torsional strength 抗扭强度
torsional stress 扭胁强,扭应力
torsional vibration 扭转振动
torsion balance 扭秤
torsion tensor 挠率张量,扭率张量
torus 螺绕环,环
total eclipse 〈天文〉全食
total internal reflection 全内反射
totality 〈天文〉全食
total momentum 总动量
total reflection 全反射
total transmittance 总透射比
tracer isotope 示踪同位素
tracer metal 痕量金属
tracer technique 示踪技术
track 径迹
track-and-guidance laser radar 跟踪制导激光雷达
track angle 径迹角,轨迹角
track chamber 径迹室
tracker 跟踪器,追踪器,跟踪装置
track fading 径迹退化,径迹衰退
tracking 探测,跟踪
tracking radar 跟踪雷达
track length 径迹长度
traction force 牵引力,拉力
tractive resistance 牵引阻力
train 系;列
train of gears 齿轮系
train of mechanism 机构系统
trajectory 轨道,轨迹;弹道;迹线
transcendental equation 超越方程
transcendental number 超越数
transconductance 跨导
transducer 传感器;换能器
transduction 转导
transfer coefficient 转移系数
transfer constant 转移常数,转移恒量
transfer curve 转移曲线
transfer factor 转移系数,转移因子
transfer function 转移函数,传递函数
transfer impedance 传递阻抗,转移阻抗
transfer matrix 转移矩阵,传递矩阵
transformer 变压器;变换器
transformer core 变压器铁芯
transience 瞬态,暂态
transient current 瞬变电流
transient equilibrium 暂态平衡,瞬变平衡
transient motion 暂态运动,瞬态运动,瞬变运动
transient pulse 瞬态脉冲
transient response 瞬态响应,瞬变响应
transient state 瞬(时状)态,暂态
transient vibration 瞬态振动,过渡振动
transistor 晶体管
transistor amplifier 晶体管放

大器
transistor circuit 晶体管电路
transistor equivalent circuit 晶体管等效电路
transistor physics 晶体管物理学
transistor-transistor logic (TTL) 晶体管-晶体管逻辑
transition 跃迁;转变
transition energy 转变能;跃迁能
transition layer 转变层,过渡层
transition loss 转变损失;渡越损失
transition matrix 跃迁矩阵
transition point 转变点
transition probability 跃迁概率,跃迁几率
transition radiation 渡越辐射,穿越辐射
transition temperature 转变温度
translation 平移,平动
translational acceleration 平动加速度,平移加速度
translational degree of freedom 平动自由度
translational energy 平动能
translational motion 平移运动
translational state 平动态
translational symmetry 平移对称
translational symmetry element 平移对称素,平移对称元
translational velocity 平动速度
translator 译码器;变换器
translucent 半透明的
translucent glass 半透明玻璃
transmission 传输;发射;透射
transmission band 传输频带,通(频)带
transmission camera 透射照相机
transmission coefficient 透射系数;传递系数
transmission electron microscope 透射电子显微镜
transmission error 传输误差
transmission experiment 透射实验
transmission film 透射膜
transmission grating 透射光栅
transmission line 传输线
transmission loss 传输损耗
transmission medium 传输媒质
transmission ratio 传动比
transmissivity 透射率
transmissometer 透射计
transmittance 透射系数,透射比
transmitter 发射机,发报机
transmitting aerial 发射天线
transmutation 嬗变,蜕变
transonic 跨声速的
transparency 透明性,透明度
transponder 应答器
transport coefficient 输运系数,迁移系数
transport equation 输运方程
transport mean free path 迁移平均自由程
transport number 迁移数
transport phenomenon 输运现象
transport theory 迁移理论,输运理论
transuranic element 铀后元素
transversal force 横力
transverse acceleration 横向加速度
transverse axis 横轴,水平轴
transverse coherence 横向相干性
transverse component 横向分量
transverse diffusion 横向扩散
transverse electric wave 横电波
transverse energy 横能量
transverse illumination 横向照明;横向照度
transverse impulse 横脉冲
transverse load 横向负载
transverse mass 横质量
transverse relaxation 横向弛豫
transverse section 横截面,横断面
transverse strength 抗挠强度,横向强度
transverse stress 横向应力
transverse translation 横向平移
transverse velocity 横向速度

transverse vibration 横振动
transverse wave 横波
transverter 换能器
trap 陷阱;陷波电路;陷俘,陷获,俘获;附着
trap circuit 陷波电路
trap model 陷阱模型
trapped ion 俘获离子
trapped mode 陷模
trapped particle 陷俘粒子,俘获粒子
trapped particle instability 陷俘粒子不稳定性,俘获粒子不稳定性
trapper 陷波器
trapping effect 陷获效应
travelling field 行移场
travelling speed 移动速率
travelling wave 行波
trebler 三倍倍频器
trial-and-error method 尝试法
triangle of vectors 矢量三角形
triangular prism 三面体棱镜,三角棱镜
triangular wave 三角形波
triboelectricity 摩擦电,摩擦起电
triboelectrification 摩擦起电
tribology 摩擦学
triboluminescence 摩擦发光
trichromatic vision 三色视觉,三原色视觉
triclinic 三斜的,三斜晶系的
triclinic crystal 三斜晶体
triclinic system 三斜晶系
tricritical point 三相临界点
triductor 三次倍频器
trigger 触发器,触发,触发电路
trigger diode 触发二极管
triggering energy 触发能
triggering pulse 触发脉冲
triggering sensitivity 触发灵敏度
trigger pulse 触发脉冲
trigger triode 触发三极管
trigonal crystal 三角系晶体
trigonal system 三角晶系
trigonometric function 三角函数
trimolecular model 三分子模型
triode 三极管
triple collision 三体碰撞
triple curve 三相曲线
triple integral 三重积分
triple point 三相点,三态点
triple star 〈天文〉三合星
triplet 三合透镜;三重态;三重线
triplet lens 三合透镜
triplet state 三重态,三重线态
triplet thin lens 三合薄透镜
tripod 三脚架
tripod magnifier 三脚放大镜
tritiated compound 氚化合物
tritide 氚化合物
tritium 氚,超重氢
triton 氚核
trivector 三维矢量
tropical year 〈天文〉回归年,分至年
troposphere 对流层
troptometer 扭转仪,扭力计,扭角仪
trough 波谷
true anomaly 〈天文〉真近点角
true error 真误差
true place 〈天文〉真位置
true value 真值
truth table 真值表
TTL (transistor-transistor logic) 晶体管-晶体管逻辑
tube 管;电子管
tube amplifier 电子管放大器
tube generator 电子管振荡发生器
tube oscillator 电子管振荡器
tube parameter 电子管参数
tunable laser 可调谐激光器
tuned amplifier 调谐放大器
tuned circuit 调谐电路
tuned grid circuit 调栅电路
tuned plate circuit 调板电路
tuner 调谐器
tuning 调谐
tuning characteristic 调谐特性
tuning coil 调谐线圈

tuning fork 音叉
tunnel current 隧道(效应)电流
tunnel diode 隧道二极管,江崎二极管
tunnel effect 隧道效应
tunnel junction 隧道结
tunnel photodiode 隧道光电二极管
tunnel transistor 隧道晶体管
turbine 涡轮,涡轮机
turbogenerator 涡轮式发电机
turbojet 涡轮喷气发动机
turbulence 湍流,紊流
turbulence intensity 湍流强度
turbulent flow 湍流,涡流,紊流
turbulent fluctuation 湍流脉动;湍流涨落现象,湍流起伏现象
turbulent flux 湍流通量
turbulent heating 湍动加热,湍流加热
turbulent jet 湍射流
turbulent medium 湍性媒质,扰动媒质
turbulent mixing 湍流混合
turbulent motion 湍动,湍流运动
turbulent region 湍流区域
turbulent resistance 湍流阻力
turbulent separation 湍流分离
turbulent shock 湍动激波,湍流激波
turbulent spectrum 湍动谱,湍流谱
Turing machine 〈计算机〉图灵机
turning point 反转点
turnmeter 转速计
turn-off 断开
turns ratio 匝数比
TV (television) 电视(机)
T violation T破坏
tweeter 高频扬声器
twin band 孪晶带
twin crystal 孪晶,双晶体
twin diode 双二极管
twin image 孪(生)像
twinkler 发光体,闪光体
twin laser 孪激光器
twin magnets 孪片磁体
twinning axis 孪晶轴,孪生轴
twin paradox 双生子佯谬
twin pulse code 双脉冲编码
twin triode 孪三极管
twist 扭曲,扭转
twist boundary 扭转晶界
twisted curve 挠曲线
twisted wire 绞合线
twisting force 扭力
twisting moment 转矩,扭矩
twisting motion 扭转运动
twisting stress 扭胁强,扭应力
twistor 扭量;磁扭线
two-beam interferometry 双光束干涉测量术,双光束干涉量度学
two-body problem 二体问题
two-component neutrino theory 二分量中微子理论
two-cycle engine 二冲程发动机
two-dimensional grating 二维光栅,平面光栅
two-photon absorption 双光子吸收
two-photon coherent state 双光子相干态
Tyndall effect 廷德尔效应
type Ⅰ superconductor 第一类超导体
type Ⅱ superconductor 第二类超导体
type-S fuse S型熔丝

Uu

U (uranium)〈化〉铀
UHF (ultrahigh frequency) 超高频,特高频
ultimate energy 极限能量
ultimate particle 基本粒子
ultimate strength 极限强度
ultracentrifuge 超速离心机
ultracrystallite 超微晶
ultrafine grain 超细颗粒
ultrahigh frequency (UHF) 超高频,特高频
ultramicroscope 超显微镜
ultrapurity 超纯度
ultrared 红外线;红外的
ultrashort laser pulse 超短激光脉冲
ultrasonic cleaning 超声清洗
ultrasonic dispersion 超声分散
ultrasonic drill 超声钻
ultrasonic energy 超声能
ultrasonic engineering 超声工程
ultrasonic erosion 超声腐蚀
ultrasonic grating 超声光栅
ultrasonic imaging 超声成像
ultrasonic inspection 超声检查,超声探伤
ultrasonic lens 超声透镜
ultrasonic light diffraction 超声致光衍射
ultrasonic luminescence 超声发光
ultrasonic nondestructive testing 超声无损检测
ultrasonic physics 超声物理学
ultrasonic processing 超声处理
ultrasonic pulse 超声脉冲
ultrasonics 超声学
ultrasonic soldering 超声焊接
ultrasonic transducer 超声换能器
ultrasonic vibration 超声振动
ultrasonic washing 超声清洗
ultrasonic wave 超声波
ultrasound 超声
ultravacuum 超真空
ultraviolet (UV) 紫外线;紫外的
ultraviolet band 紫外谱带,紫外区,紫外波段
ultraviolet catastrophe 紫外灾难
ultraviolet detector 紫外线探测器
ultraviolet dosimeter 紫外线剂量计
ultraviolet irradiation 紫外线辐照
ultraviolet lamp 紫外线灯
ultraviolet microscope 紫外线显微镜
ultraviolet photon 紫外光子
ultraviolet radiation 紫外辐射
ultraviolet ray 紫外线
ultraviolet spectrogram 紫外光谱图
ultraviolet spectrograph 紫外摄谱仪
umbra 本影,暗影
unactivated state 未激活态
unbound state 非束缚态
uncertainty principle 测不准原理,不确定原理
undercorrected lens 欠校正透镜,校正不完全透镜
undercoupling 欠耦合,耦合不足
underdamping 欠阻尼,弱阻尼

underwater photography 水下摄影术
underwater probing 水下探测
underwater sound projector 水下声标，声呐发射器
underwater vision 水下视觉
undistorted image 无畸变图像，不失真图像
uniaxial crystal 单轴晶体
unified-field theory 统一场论
uniform acceleration 匀加速度
uniform circular motion 匀速圆周运动
uniform dielectric 均匀电介质
uniform illumination 均匀照明
uniformity of spacetime 时空均匀性
uniform motion 匀速运动
uniform speed 匀速，等速
uniform velocity 匀速
unijunction transistor 单结晶体管
unilateral conduction 单向导电，单向导电
unimodal laser 单模激光器，单峰激光器
unimodular transformation 单模变换，幺模变换
union 并集
unipolar transistor 单极晶体管
uniqueness theorem 唯一性定理，单值定理
unit 单位；单元
unitarity 幺正性
unitary group 幺正群，酉群
unitary matrix 幺正矩阵，酉矩阵
unitary operator 幺正算符，酉算子
unitary spin 幺旋
unitary transformation 幺正变换，酉变换
unit cell 晶胞，单胞
unit cell dimension 晶胞大小
unit charge 单位电荷
uniterminal axis 单极轴
unit magnetic pole 单位磁极
unit mass 单位质量
uniton 单子
unit vector 单位矢量，单位向量
universal constant 普适常量，普适恒量
universal gas constant 普适气体常量
universal gravitation 万有引力
universality 普适性
universal meter 多用电表，万用表
universal motor 交直流两用电动机，通用式电动机
universal resource locator (**URL**) 〈计算机〉统一资源定位符
universe 宇宙
unpolarized 非偏振的，非极化的
unpolarized light 非偏振光
unsaturated steam 不饱和水蒸气
unsaturated vapour 未饱和汽
unstable equilibrium 不稳定平衡
unsteady constraint 非定常约束
unsteady state 非稳态
upper atmosphere 高层大气
upper bound 上界
up quark 上夸克
uranium (**U**) 〈化〉铀
uranium series 铀系
URL (**universal resource locator**) 〈计算机〉统一资源定位符
U-tube U形管
UV (**ultraviolet**) 紫外线；紫外的
uvaser 远紫外激光器
UV laser 紫外线激光器

Vv

V (volt) 伏(特)(电压单位)
vacancy 空位,空格点,缺位
vacancy cluster 空位团,空位群
vacuometer 真空计
vacuscope 真空仪,真空计
vacuum 真空
vacuum box 真空盒
vacuum brake 真空闸,真空制动
vacuum breaker 真空断流器
vacuum chamber 真空室
vacuum circuit-breaker 真空断流器
vacuum cleaner 真空吸尘器,真空清洁器
vacuum coating 真空镀膜,真空涂敷
vacuum coefficient 真空系数
vacuum cooling 真空冷却
vacuum dilatometer 真空膨胀计
vacuum discharge 真空放电
vacuum distillation 真空蒸馏
vacuum expander 真空膨胀器
vacuum extraction 真空萃取,真空提取
vacuum field 真空场
vacuum firing 真空点火
vacuum flask 真空(保温)瓶,杜瓦瓶
vacuum gauge 真空规,真空计
vacuum grown crystal 真空长成晶体
vacuum indicator 真空度指示器
vacuum insulation 真空绝缘
vacuum intensifier 真空增强器
vacuum interrupter 真空断续器
vacuum lamp 真空灯
vacuum leak 真空泄漏,真空损失,漏气
vacuum manometer 真空压力计
vacuum measurement 真空测量
vacuum melting 真空熔炼
vacuum of space 空间真空,(宇宙)空间真空度
vacuum photocell 真空光电管
vacuum phototube 真空光电管
vacuum physics 真空物理学
vacuum plant 真空装置
vacuum pump 真空泵,蒸汽吸水机
vacuum purification 真空净化,真空提纯
vacuum space 真空室,真空空间
vacuum state 真空态
vacuum switch 真空开关
vacuum thermocouple 真空热电偶,真空温差电偶
vacuum thermopile 真空热电堆,真空温差电堆
vacuum tower 真空蒸馏塔
vacuum trap 真空阱,真空聚集槽
vacuum tube 真空管,电子管
vacuum tweezer 真空钳,真空镊子,真空机械手
vacuum valve 真空管,电子管;真空阀;真空整流管
vacuum vessel 真空瓶,真空容器
valence 〈化〉(化合)价,原子价
valence band 价(电子)带
valence electron 价电子
valence force (原子)价力
valence nucleon 价核子
valence shell 价电子层,价壳层
valence state 价态
valency 〈化〉(化合)价,原子价

valency charge 价电荷
valve 阀;真空管,电子管
Van Allen belt 辐射带,范艾伦带
Van de Graaff accelerator 范德格拉夫加速器,静电加速器
Van de Graaff generator 范德格拉夫起电机
van der Waals' equation 范德瓦耳斯方程
van der Waals' force 范德瓦耳斯力
vane anemometer 叶轮风速计
vaporization 汽化
vapour 蒸汽,汽
vapour density 蒸汽密度,水汽密度
vapour-deposition method 汽相沉积法
vapour phase 汽相,蒸汽相
vapour pressure 蒸汽压
vapour pump 蒸汽泵
vapour tension 蒸汽压,汽压
varactor 变容二极管;可变电抗器
variable acceleration 变加速度
variable autotransformer 可调自耦变压器
variable condenser 可变电容器
variable filter 可变滤光片,可变滤波器
variable focal length 变焦(距)
variable inductance 可变电感
variable inductor 可变电感器
variable load 可变负载,可变负荷
variable mass system 变质量系
variable motion 变速运动
variable reluctance 可变磁阻
variable resistor 可变电阻器
variable speed 变速,可变速度
variable star 〈天文〉变星
variable velocity 变速,可变速度
variance 方差
varifocal lens 可变焦距镜头;变焦透镜
varifocal objective 变焦物镜
variocoupler 可变耦合器
variode 变容二极管
variodenser 可变电容器
variometer 可变电感器,变感器,磁变仪
variopter 滑动光学计算尺
varistor 变阻器
varitron 变换子
VDU (visual-display unit) 视频显示装置
vector 矢量,向量
vector analysis 矢量分析
vector current 矢量流
vector density 向量密度,矢量密度
vector diagram 矢量图;谐量图
vector field 矢量场
vector flux 矢通量,向量流量
vector meson 矢量介子,矢性介子
vector meter 矢量计
vector product 向量积,叉积,矢(量)积
vector quantity 矢量,向量
vector space 向量空间
velocimeter 速度计,速度表
velocity 速度
velocity field 速度场
velocity modulation 调速
velocity of following 牵连速度
velocity of light 光速
velocity of migration 徙动速度
velocity ratio 速(度)比
velocity space 速度空间
Venn diagram 维恩图解
Venturi tube 文氏管,文丘里管
vernier 游标,游标尺
vernier caliper 游标卡尺
vernier condenser 微变电容器
vernier device 微调装置
vernier focusing 微变聚焦
vertex angle 顶角
vertex of mirror 镜顶
vertical incidence 垂直入射,正入射,法向入射
vertical line 垂直线,法线,铅垂线

vertical projection 垂直投影
very high frequency (VHF) 甚高频
very-large-scale integrated circuit (VLSI circuit) 超大规模集成电路
very low frequency (VLF) 甚低频
VHF (very high frequency) 甚高频
vibration 振动,振荡
vibrational band 振动能带,振动谱带
vibrational excitation 振动激发,振动激励
videotape 录像带
vidicon 光导摄像管,视像管
viewing field 视场
viewing lens 观测物镜
view stop 视场光阑,场阑
vignetting stop 渐晕光阑,格晕光阑,护真光阑
virial equation 维里方程
virtual displacement 虚位移
virtual image 虚像
virtual particle 虚粒子
virtual state 虚态
virtual work 虚功
viscoelasticity 黏弹性
viscometer 黏度计
viscosity 黏性,黏滞性,黏滞度
viscosity index 黏度指数
viscous fluid 黏性流体,黏滞流体
viscous friction 黏滞摩擦
visibility 可见度,视见度,能见度
visible light 可见光
visible range 可见(光谱)区
visible spectrum 可见光谱
vision 视觉,视力
vision function 视觉函数,视见函数
visual binary 目视双星
visual brightness 视觉亮度
visual diameter 视直径
visual-display unit (VDU) 视频显示装置
visual photometer 目测光度计
VLF (very low frequency) 甚低频
VLSI circuit (very-large-scale integrated circuit) 超大规模集成电路
volometer 伏安表
volt (V) 伏(特)(电压单位)
voltage 电压,伏特数
voltage amplifier 电压放大器
voltage comparator 电压比较器
voltage divider 分压器,电压分配器
voltage stabilizing diode 稳压二极管
voltaic battery 伏打电池
voltaic cell 伏打电池,原电池
voltaic electricity 伏打电
voltaic pile 伏打电堆
voltaism 流电;流电学
voltameter 电量计,库仑计
volt-ampere 伏安,伏特安培
volt-ampere characteristics 伏安特性曲线
voltmeter 伏特计,电压表
voltmeter-ammeter method 伏安法
voltolization 无声放电处理;高电压处理,电聚
volume 体积;容量;音量
volume charge 体电荷
volume expansion 体积膨胀
volume force 体积力
volume integral 体积分
volume level 强度级,响度级
volumenometer 体积计
volumeter 体积计,容量计
volume velocity 体积速度
vortex 涡旋
vortex axis 旋涡轴
vortex flow 涡流
vortex pair 涡(线)对,旋涡偶
vortex stabilization 涡旋稳定
vortex strip 涡流带
vorticity 涡度,涡旋强度

Ww

W (watt) 瓦(特)(电功率单位)
wafer 晶片;干胶片
waist radius (光)束腰半径
wall effect (器)壁效应
water cooling 水冷却
water engine 水压机,水力发动机
water level 水位,水平面;水准仪
water level gauge 水准仪,水位表
water line 吃水线,水位线
water meter 水表
water motor 水力发动机
water power 水力,水能
water pump 水泵,抽水机
water softener 软水剂
water vapour 水蒸气
water wave 水波
watt (W) 瓦(特)(电功率单位)
wattage 瓦特数
watthour 瓦特时
watthour meter 瓦时计,电度表
wattmeter 瓦特计
wave 波
wave amplitude 波幅,波度
wave analysis 波形分析
wave band 波段
wave crest 波峰,波顶
wave damping 波阻尼
wave detector 检波器
wave director 导波体,波导
wave equation 波动方程
wave field 波场
wave form 波形
wave formalism 波论
wave front 波前,波阵面
wave function 波函数
waveguide 波导;波导管
wavelength 波长
wavelength spectrometer 波长分光计
wave loop 波腹
wave mechanics 波动力学
wavemeter 波长计,波频计
wave number 波数
wave optics 波动光学
wave packet 波包
wave-particle duality 波粒二象性
wave pattern 波型,波动图式
wave propagation 波的传播
wave radiation 波辐射
wave range 波段
wave reflection 波反射
wave refraction 波折射
wave resistance 波阻
wave shape 波形
wavestrip 波带
wave tail 波尾
wave theory 波动说,波动理论
wave trough 波谷
wave vector 波矢(量)
wave velocity 波速(度)
wavicle 波粒二象性
Wb (weber) 韦伯(磁通量单位)
W-boson W 玻色子
weak coupling 弱耦合
weak interaction 弱相互作用
Web 环球网,万维网
weber (Wb) 韦伯(磁通量单位)
wedge shape 楔形(断面)
weight 重量
weightlessness 失重
Weinberg-Salam model (WS model) 温伯格-萨拉姆模型,

WS模型
Weston cell 韦斯顿(标准)电池
wet-and-dry bulb hygrometer 干湿球湿度计
Wheatstone bridge 惠斯通电桥,单臂电桥
wheel 轮,滑轮;叶片
wheel and axle 轮轴
white dwarf 〈天文〉白矮星
white hole 白洞
white light 白光
white noise 白噪声
white spectrum 白谱
wide-angle lens 广角镜头,广角透镜
wide-aperture lens 大孔径透镜
wideband 宽(频)带
Wiedemann-Franz law 维德曼-弗兰兹定律
Wien formula 维恩公式
Wien's displacement law 维恩位移(定)律
Wigner energy 维格纳能量
Wigner nuclides 维格纳核素
Wigner-Witmer rules 维格纳-威特默定则
Wilson cloud chamber 威尔逊云室
Wimshurst machine 维姆胡斯特起电机
wind energy 风能
wind gauge 风力计,风速计
winding 绕组;绕法
window 窗
wind power 风力
wire 导线,电线
wire grating 线(光)栅
wireless telegraphy 无线电报(学)
wireless telephony 无线电话学
Wollaston prism 沃拉斯顿棱镜
Wood's metal 伍德合金
woofer 低频扬声器
word 〈计算机〉字
work 功
work function 功函数,逸出功
work-hardening 加工硬化,功致硬化
world line 世界线
World Wide Web (WWW) 环球网,万维网
W particle W粒子
wrought iron 熟铁,锻铁
WS model (Weinberg-Salam model) 温伯格-萨拉姆模型,WS模型
WWW (World Wide Web) 环球网,万维网

Xx

xaser X 射线激光器,X 射线激射器;X 射线激光
xerography 静电复印
X-ray X 射线,X 光
X-ray astronomy X 射线天文学
X-ray crystallography X 射线晶体学
X-ray diffraction X 射线衍射
X-ray fluorescence X 射线荧光
X-ray laser X 射线激光器,X 射线激射器;X 射线激光
X-ray spectrum X 射线谱
X-ray tube X 射线管

Y y

Y (**yttrium**) 钇
Yagi aerial 八木天线,引向天线
YAG laser YAG 激光器,钇铝石榴石激光器
Yang-Mills field 杨(振宁)-米尔斯场,杨(振宁)-米尔斯规范场
yard 码(长度单位)
Yb (**ytterbium**) 镱
year 年(时间单位)
yield 屈服;产额
yield point 屈服点
Young modulus of elasticity 杨氏弹性模量
Young's double-slit experiment 杨氏双缝实验
Young's slits 杨氏双缝
ytterbium (**Yb**) 镱
yttrium (**Y**) 钇
Yukawa interaction 汤川相互作用

Zz

Z boson Z玻色子
Zeeman effect 塞曼效应
Zeeman splitting 塞曼(谱线)分裂
Zener breakdown 齐纳击穿
Zener diode 齐纳二极管,稳压二极管
zenith 顶点;〈天文〉天顶
zero adjuster 零点校正器
zero adjustment 零点调整
zero beat 零拍
zero graduation 零刻度
zero-point energy 零点能
zero reading 零读数,起点读数
zeroth law of thermodynamics 热力学第零定律
Zn (zinc) 锌
zodiac 〈天文〉黄道,黄道带
zodiacal light 黄道光,黄道带光
zone axis 晶带轴
zone index 晶带指数
zone plate 波带片,波带板
zone refining 逐区精炼,区域精炼,区域提纯
zoom lens 变焦镜头,变焦透镜
zoom ratio (可)变焦比
Z particle Z粒子
zwitterion 两性离子

非英文字母开头的词条

α-decay α衰变
α-radioactivity α放射性
α-ray α射线
β-decay β衰变
β-radioactivity β放射性
β-ray β射线
γ-decay γ衰变
γ-radioactivity γ放射性
γ-ray γ射线
θ vacuum θ真空
μA（microampere） 微安（电流强度单位）
μCi（microcurie） 微居里（放射性活度单位）
μm（micron） 微米（长度单位）
π（pi） 圆周率
Ω（ohm） 欧（姆）（电阻单位）

汉英部分

Aa

阿贝成像原理　Abbe principle of image formation
阿贝尔群　Abelian collection; Abelian group
阿贝双衍射原理　Abbe double-diffraction principle
阿德勒反常　Adler anomaly
阿伏伽德罗常量　Avogadro constant
阿伏伽德罗定律　Avogadro's law
阿根图　Argand diagram
阿基米德原理　Archimedes principle
阿普尔顿层　Appleton layer
阿特伍德机　Atwood machine
锕系　actinium series
埃　angstrom
锿　Es (einsteinium)
矮星　dwarf star
艾里斑　Airy disk
爱迪生电池　Edison cell
爱因斯坦场方程　Einstein field equation
爱因斯坦-德哈斯效应　Einstein-de Haas effect
爱因斯坦等效原理　Einstein equivalence principle
爱因斯坦方程　Einstein equation
爱因斯坦关系　Einstein relation
爱因斯坦红移　Einstein redshift; gravitational redshift
爱因斯坦-麦克斯韦方程　Einstein-Maxwell equation
爱因斯坦模型　Einstein model
爱因斯坦频率　Einstein frequency
爱因斯坦曲率张量　Einstein curvature tensor
爱因斯坦位移　Einstein shift
爱因斯坦温度　Einstein temperature
爱因斯坦系数　Einstein coefficients
安伏欧计　avometer
安培　A (ampere)
安培表　ampere-meter
安培定律　Ampère's law
安培定则　Ampère's rule
安培计　ammeter
安培力　Ampère force
安培小时　ampere-hour
安培匝数　ampere-turn
安全玻璃　safety glass
安全灯　safelight; safety lamp
安全阀　safety valve
安全离合器　safety clutch
安全条件　safety condition
安全因数　safety factor

安全因子 safety factor
安时 ampere-hour
安匝(数) ampere-turn
氨分子钟 ammonia clock
鞍点 saddle point
暗场 dark field
暗带 dark band
暗视场 dark field
暗物质 dark matter
暗线 dark line
暗影 umbra
暗噪声 dark noise
盎司 oz (ounce)
凹面的 concave
凹面光栅 concave grating
凹面镜 concave mirror; diverging mirror
凹透镜 concave lens; diverging lens
凹凸透镜 concave-convex lens
凹陷 inflection
凹形的 concave
奥 Oe (oersted)
奥伯斯佯谬 Olbers' paradox
奥氏体 austenite
奥斯特 Oe (oersted)
奥斯特瓦尔德催热 Ostwald ripening
奥斯特瓦尔德稀释定律 Ostwald's dilution law
奥托发动机 Otto engine

Bb

八比特组 octet
八度音阶 octave
八进位法 octal notation
八进制计数法 octal notation
八木天线 Yagi aerial
八位(位)组 octet
巴 bar
巴比涅补偿器 Babinet compensator
巴比涅原理 Babinet principle
巴比特合金 Babbitt metal
巴丁-肖克莱公式 Bardeen-Shockley formula
巴耳末(谱线)系 Balmer series
巴克豪森效应 Barkhausen effect
靶 object; target
靶电流 target current
靶(恩) b (barn)
靶核 target nucleus
靶理论 target theory
靶粒子 target particle
靶室 target chamber
靶同位素 target isotope
靶学说 target theory
白矮星 white dwarf
白炽灯 filament lamp; incandescent lamp
白洞 white hole
白光 white light
白镴 pewter
白谱 white spectrum
白热 incandescence
白霜 hoarfrost
白噪声 white noise
白昼视觉 photopic vision
百分标 centigrade scale
百分刻度 centigrade scale
摆 pendulum
摆动 oscillation
摆式计数器 pendulum meter
摆线 cycloid
摆运动 pendulum motion
板(极) plate
板形整流器 plate rectifier
半波片 half-wave plate
半波损失 half-wave loss
半波天线 half-wave aerial
半波整流器 half-wave rectifier
半导体 semiconductor
半导体二极管 semiconductor diode
半导体激光器 semiconductor laser
半导体结 semiconductor junction
半导体物理学 semiconductor physics
半导体元件 semiconductor element
半导性 semiconductivity
半电池 half cell
半定量的 semiquantitative
半镀银镜 half-silvered mirror
半对称共振腔 half-symmetric resonator
半反射器 semireflector
半峰半宽 half width at half maximum
半峰全宽 half-peak width
半经典方法 semiclassical approach
半经典近似 semiclassical approximation
半经验的 semiempirical

半晶态 semicrystalline state
半宽度 half-width；semibreadth
半衰期 half-life；half-time
半透明玻璃 translucent glass
半透明的 translucent
半透膜 semipermeable membrane
半影 penumbra
半影器件 half-shade device
半影装置 half-shade device
半值厚度 half-value thickness
半周期带 half-period zone
伴波 satellite wave
伴峰 satellite peak
伴线 satellite line
伴线结构 satellite structure
伴星 companion star
伴星系 companion galaxy
傍轴成像 paraxial imagery
傍轴的 paraxial
傍轴方程 paraxial equation
傍轴光线 paraxial ray
傍轴焦点 paraxial focus
傍轴近似 paraxial approximation
傍轴区 paraxial region
傍轴条件 paraxial condition
磅 pound
磅达 poundal
包覆涂层 cladding
饱和 saturation
饱和电流 saturation current
饱和电阻 saturation resistance
饱和度 saturation
饱和光谱学 saturation spectroscopy
饱和极化强度 saturated polarization
饱和器 saturator
饱和强度 saturation intensity
饱和区 saturation region
饱和曲线 saturation curve
饱和溶液 saturated solution
饱和增益 saturated gain
饱和蒸汽 saturated vapour
饱和值 saturation (value)
保护气体 blanket gas
保健物理学 health physics
保角变换 conformal transformation
保守场 conservative field
保守力 conservative force
保守系 conservative system
保险丝 fuse
保险丝盒 fuse box
保真度 fidelity
保真性 fidelity
暴缩 implosion
暴胀宇宙 inflationary universe
曝光 exposure
曝光表 exposure meter
曝光量 exposure
爆发星系 explosive galaxy
爆沸 bumping
爆击 knocking
爆聚 implosion
爆聚加热 implosion heating
爆鸣气 detonating gas；electrolytic gas
爆炸波 blast wave
爆炸量热器 bomb calorimeter
爆震音 knocking
杯式(水银)气压计 cistern barometer
北极 N pole
北极光 aurora borealis
贝尔 bel
贝尔定理 Bell's theorem
贝壳状断裂 conchoidal fracture
贝壳状裂痕 conchoidal fracture
贝可(勒尔) Bq (becquerel)
贝克曼(水银)温度计 Beckmann thermometer
贝纳胞 Bénard cell
贝塞尔函数 Bessel function；cylindrical harmonics
贝塞麦炼钢法 Bessemer process
贝陀立式化合物 Berthollide compound
贝叶斯定理 Bayes theorem
备份 backup
备用元件 backup

背景辐射 background radiation
倍频 frequency doubling
倍频程 octave
倍频程滤波器 octave filter
倍频带 octave band
倍增器 multiplier
倍增因数 multiplication factor
被动器件 passive device
被积函数 integrand
被吸附物 adsorbate
本初子午圈 prime meridian
本初子午线 prime meridian
本底辐射 background radiation
本底噪声 noise background
本构方程 constitutive equation
本机振荡器 local oscillator
本轮 epicycle
本生灯 Bunsen burner
本生电池 Bunsen cell
本体模量 bulk modulus
本星系群 Local Group
本影 umbra
本征半导体 intrinsic semiconductor
本征方程 intrinsic equation
本征函数 eigenfunction; proper function
本征模 eigenmode
本征能 intrinsic energy
本征频率 eigenfrequency
本征矢量 eigenvector
本征态 characteristic state; eigenstate
本征载流子 intrinsic carrier
本征振荡 eigen oscillation
本征振动 eigenvibration
本征值 eigenvalue; proper value
崩沸 bumping
泵 pump
比电离度 specific ionization
比(反应)面积 specific surface
比放射性 specific activity
比荷 specific charge
比活度 specific activity
比较半衰期 comparative half-life
比较光谱 comparison spectrum
比结合能 specific binding energy
比例 proportion
比例尺 scale
比例绘图器 pantograph
比例极限 proportional limit
比面积 specific surface
比热 specific heat
比热反常 specific heat anomaly
比热容 specific heat capacity
比容 specific volume
比色法 colorimetry
比色高温计 colorimetric pyrometer
比色计 chromometer
比湿 specific humidity
比释动能 kerma
比特 bit
比特粉纹图样 Bitter powder pattern
比体积 specific volume
比旋 specific rotation
比影光度计 shadow photometer
比重 specific gravity; specific weight
比重计 hydrometer
比阻 specific resistance
毕奥-萨伐尔定律 Biot-Savart law
闭共振腔 closed resonator
闭合电路 closed circuit
闭合共振器 closed resonator
闭路 closed circuit
闭系 closed system
闭宇宙 closed universe
壁效应 wall effect
避雷线 lightning conductor
避雷针 lightning rod
避雷装置 lightning conductor
边际效应 edge effect
边界层 boundary layer
边界散射 boundary scattering
边界条件 boundary condition
边频带 sideband
边缘 fringe

边缘发射 edge emission
边缘效应 edge effect
边缘增强(效应) edge enhancement
边值关系 boundary relation
边值问题 boundary-value problem
编译程序 compiler
变比化合物 Berthollide compound
变感器 variometer
变换器 transformer；translator
变换谐波量分析 Fourier analysis
变换因数 conversion factor
变换子 varitron
变加速度 variable acceleration
变焦比 zoom ratio
变焦镜头 zoom lens
变焦(距) variable focal length
变焦透镜 varifocal lens；zoom lens
变焦物镜 varifocal objective
变流机 converter
变流器 converter
变流因数 conversion factor
变频管 converter
变频器 converter
变容二极管 varactor；variode
变色 chromatism
变速 variable speed；variable velocity
变速运动 non-uniform motion；variable motion
变相器 phase changer
变星 variable star
变形光学系统 anamorphic optical system
变形核 deformed nucleus
变形镜头 anamorphote lens
变形壳模型 deformed shell model
变旋现象 mutarotation
变压器 transformer
变压器铁芯 transformer core
变异旋光 multi-rotation；mutarotation
变质量系 variable mass system
变阻器 rheostat；varistor
遍举过程 exclusive process
遍历假说 ergodic hypothesis
遍历性 ergodicity
标称功率 nominal power
标称偏差 nominal deviation
标乘法 scalar multiplication
标度 scale
标度变换 scale transformation
标度参数 scale parameter
标度律 scaling law
标度无关变量 scaling variable
标度无关律 scaling law
标度值 scale value
标记 labelling
标积 scalar product
标量 scalar (quantity)
标量场 scalar field
标量电子 scalar electron；selectron
标量积 scalar product
标量介子 scalar meson
标量介子理论 scalar meson theory
标量矩阵 scalar matrix
标量夸克 squark
标量偶理论 scalar pair theory
标量轻子 slepton
标量通量 scalar flux
标量相互作用 scalar interaction
标势 scalar potential
标志光谱 characteristic spectrum
标准大气压 standard atmospheric pressure
标准电池 normal cell；standard cell
标准电极 standard electrode
标准电极电位 standard electrode potential
标准模型 standard model
标准偏差 standard deviation
标准态 standard state
标准温度和压力 s. t. p. (standard temperature and

pressure)
标准误差 standard error
标准型 standard form
标准钟 standard clock
标准状态 normal state
表观膨胀 apparent expansion
表观深度 apparent depth
表观质量 apparent mass
表面波 surface wave
表面淬火 surface hardening
表面耗尽层 surface depletion layer
表面活性剂 surface active agent; surfactant
表面积分 surface integral
表面结构 surface structure
表面流 surface current
表面能 surface energy
表面熵 surface entropy
表面态 surface state
表面物理学 surface physics
表面应力 surface stress
表面硬化 case hardening; surface hardening
表面张力 surface tension
表面张力波 capillary wave
表面振动 surface vibration
冰 ice
冰点 ice point
冰点降低常数 cryoscopic constant
冰晶 ice crystal
冰模型 ice model
冰熵 ice entropy
冰式晶体 ice crystal
冰条件 ice condition
冰盐 cryohydrate
饼形图 pie chart
禀性方程 intrinsic equation
并-串联电路 parallel-series circuit
并合原理 combination principle
并激绕法 shunt winding
并激绕组 shunt winding
并集 union
并联 connection in parallel
并联电路 parallel circuits
并联共振 parallel resonance
并协原理 complementary principle
并行处理 parallel processing
波 wave
波包 wave packet
波长 wavelength
波长分光计 wavelength spectrometer
波长计 wavemeter
波场 wave field
波带 band; wavestrip
波带板 zone plate
波带片 zone plate
波导 wave director; waveguide
波导管 waveguide
波的传播 wave propagation
波顶 wave crest
波动 fluctuation; ripple
波动方程 wave equation
波动光学 wave optics
波动理论 wave theory
波动力学 wave mechanics
波动说 wave theory
波动图式 wave pattern
波度 wave amplitude
波段 wave band; wave range
波反射 wave reflection
波峰 peak; (wave) crest
波幅 wave amplitude
波幅畸变 amplitude distortion
波辐射 wave radiation
波腹 antinode; wave loop
波谷 (wave) trough
波函数 wave function
波节 node
波粒二象性 wave-particle duality; wavicle
波论 wave formalism
波模简并度 mode degeneracy
波模锁定 mode-locking
波频计 wavemeter
波谱 spectrum
波谱学 spectroscopy
波前 wave front

波矢(量) wave vector
波束角 beam angle
波数 wave number
波速(度) wave velocity
波特 Bd (baud)
波尾 wave tail
波纹(度) ripple
波形 wave form; wave shape
波形分析 wave analysis
波形因数 form factor
波型 mode; wave pattern
波型耦合 mode coupling
波型图样 mode pattern
波折射 wave refraction
波阵面 wave front
波阻 wave resistance
波阻抗 surge impedance
波阻尼 wave damping
玻恩-奥本海默近似 Born-Oppenheimer approximation
玻恩-格林方程 Born-Green equation
玻恩-哈伯循环 Born-Haber cycle
玻恩-迈耶势 Born-Mayer potential
玻尔半径 Bohr radius
玻尔磁子 Bohr magneton
玻尔理论 Bohr theory
玻尔频率条件 Bohr frequency condition
玻尔学说 Bohr theory
玻尔原子模型 Bohr atom model
玻尔兹曼常量 Boltzmann constant
玻尔兹曼方程 Boltzmann equation
玻尔兹曼分布律 Boltzmann distribution law
玻尔兹曼因子 Boltzmann factor
玻璃 glass
玻璃化转变 glass transition
玻璃化转变温度 glass transition temperature
玻璃纤维 glass fibre
玻璃液体温度计 liquid-in-glass thermometer
玻璃陨石 tektite
玻色-爱因斯坦凝聚 Bose-Einstein condensation
玻色-爱因斯坦统计法 Bose-Einstein statistics
玻色子 boson
玻色(子)弦 bosonic string
玻意耳定律 Boyle's law
伯努利定理 Bernoulli theorem
伯努利方程 Bernoulli equation
泊 poise
泊松比 Poisson's ratio
泊松方程 Poisson's equation
泊松分布 Poisson distribution
泊松括号 Poisson bracket
泊肃叶定律 Poiseuille law
薄膜 membrane
薄膜沸腾 film boiling
薄膜干涉 (thin) film interference
薄膜光学 thin film optics
薄透镜 thin lens
补偿 offset
补偿板 compensating plate
补偿电流 offset current
补偿目镜 compensating eyepiece; offset eyepiece
补色 complementary colour
不摆 deadbeat
不饱和水蒸气 unsaturated steam
不闭合导线 open traverse
不带电的 neutral
不带电粒子 chargeless particle
不等 inequality
不等边三角形 scalene
不等式 inequality
不定积分 indefinite integral
不动点 fixed point
不对称裂变 asymmetrical fission
不对称原子 asymmetric atom
不规则反射 irregular reflection
不规则三角形 scalene
不规则振荡 hunting
不尽根 surd
不可逆过程 irreversible process

不可逆性 irreversibility
不可压缩性 incompressibility
不良导体 poor conductor
不确定性 indeterminacy
不确定原理 principle of indeterminacy；uncertainty principle
不失真图像 undistorted image
不碎玻璃 safety glass
不透明度 opacity
不透明性 opacity
不稳定平衡 unstable equilibrium
不稳定性 instability
不相交定则 noncrossing rule
不相容性 incompatibility
不相容原理 exclusion principle
不锈钢 stainless steel
不晕镜 aplanat
不胀钢 Invar
布尔代数 Boolean algebra
布尔登压力计 Bourdon gauge
布居反转 population inversion
布拉格定律 Bragg's law
布拉格矢量 Bragg vector
布拉格衍射 Bragg diffraction
布拉开系 Brackett series
布拉维格 Bravais lattice
布拉维群 Bravais group
布朗运动 Brownian motion
布里渊函数 Brillouin function
布里渊区 Brillouin zone
布里渊散射 Brillouin scattering
布洛赫波函数 Bloch wave function
布洛赫定理 Bloch's theorem
布儒斯特窗 Brewster window
布儒斯特定律 Brewster's law
布儒斯特角 Brewster angle
布氏硬度 Brinell hardness
步进电动机 step motor
步进开关 stepping switch
步进马达 step motor
步进衰减器 step attenuator
步进调节器 step regulator
步进制电子管 stepping tube；step tube
钚 Pu（plutonium）
部分偏振光 partial polarized light
部分相干光 partially coherent light
部分子 parton
部件 component

Cc

擦边角动量 grazing angular momentum
擦边碰撞 grazing collision
材料科学 material science
材料用量 amount of substance
采样器 sampler
彩色电视 colour television
彩色全息术 colour holography
彩色照相术 colour photography
参变激发 parametric excitation
参变荧光 parametric fluorescence
参考光束 reference beam
参考系 frame of reference; reference frame; reference system
参量 parameter
参量放大器 parametric amplifier
参量激励 parametric excitation
参量空间 parameter space
参数 parameter
参数变换 parameter transformation
参数方程 parametric equation
参数曲线 parameter curve
参照系 frame of reference; reference frame; reference system
残留图像 residual image
粲 charm
粲夸克 c-quark
粲数 charm
槽式气压计 cistern barometer
槽纹光栅 graticule
测不准原理 principle of indeterminacy; uncertainty principle
测地线 geodesic (line)
测地学 geodesy
测高计 altimeter; cathetometer; hypsometer
测光 photometry
测节器 nodal slide
测距仪 telegauge; telemeter
测力计 dynamometer
测量 measurement
测量误差 measurement error
测量仪表 meter
测气管 eudiometer
测声计 acoustimeter
测微分光光度计 microspectrophotometer
测微光度计 microdensitometer; microphotometer; photomicrometer
测微计 micrometer
测微密度计 microdensitometer
测微目镜 micrometer eyepiece
测温三角锥 pyrometric cone
测温性质 thermometric property
测远计 telegauge
测震仪 seismograph
策动力 driving force
层流 laminar flow; parallel flow
层析X射线照相术 tomography
层析成像 tomography
层状结构 layer structure
层状相 smectic phase
层子 straton
叉积 cross product; vector product
差动齿轮系 differential gear

train
差分放大器 differential amplifier
差拍 beats
差频 difference frequency
差示温度计 differential thermometer
插层 intercalation
插孔 jack
插口 socket
插值多项式 interpolation polynomial
插值函数 interpolation function
插座 jack；socket
查理定律 Charles' law
柴油机 diesel engine
掺杂 doping
产额 yield
产生时间 generation time
长波 long wave
长程力 long-range force
长度 length
长度收缩 length contraction
长焦镜头 telelens；telephoto lens
长期平衡 secular equilibrium
长途电信 telecommunication
长途通讯 telecommunication
尝试法 trial-and-error method
常规电流 conventional current
常量 constant
常数 constant
常温常压 N. T. P.（normal temperature and pressure）
常用对数 common logarithm
场 field
场磁铁 field magnet
场镜 field lens
场阑 field stop；view stop
场离子显微镜 FIM（field-ionization microscope；field-ion microscope）
场论 field theory
场强计 electrometer of field strength
场曲 curvature of field
场效应 field effect
场效应晶体管 FET（field-effect transistor）
场效应器件 FED（field-effect device）
场源 field source
场致发光 electroluminescence
场致发射 field emission
场致发射显微镜 FEM（field-emission microscope）
场致荷电 field charging
超纯度 ultrapurity
超大规模集成电路 VLSI circuit（very-large-scale integrated circuit）
超大气压 superpressure
超导磁浮 superconducting magnetic levitation
超导磁体 superconducting magnet
超导磁铁 superconducting magnet
超导的 superconducting
超导(电)性 superconductivity
超导膜 superconducting film
超导体 superconductor
超点阵 superlattice
超电势 overpotential
超电位 overpotential
超短波辐射 hard radiation
超短激光脉冲 ultrashort laser pulse
超对称(性） supersymmetry
超额电子 excess electron
超辐射 superradiance；superradiation
超高能质子 superproton
超高频 SHF（super high frequency）；UHF（ultrahigh frequency）
超高频振荡三极管 dyotron
超格点 superlattice
超功率激光器 superpower laser
超荷 hypercharge
超混沌 hyperchaos
超晶格 superlattice

超精细分裂 hyperfine splitting
超精细结构 hyperfine structure
超巨星 supergiant
超距作用 action at a distance
超空间 superspace
超临界的 supercritical
超临界状态 above-critical state
超流(动)性 superfluidity
超流态 superfluidity
超流体 superfluid
超流性 superfluidity
超密恒星 superdense star
超热中子 above-thermal neutron; epithermal neutron
超容许跃迁 superallowed transition
超弱相互作用 superweak interaction
超声 ultrasound
超声波 supersonic wave; ultrasonic wave
超声波的 supersonic
超声波发生器 supersonic generator
超声成像 ultrasonic imaging
超声处理 ultrasonic processing
超声发光 ultrasonic luminescence
超声分散 ultrasonic dispersion
超声腐蚀 ultrasonic erosion
超声工程 ultrasonic engineering
超声光栅 ultrasonic grating
超声焊接 ultrasonic soldering
超声换能器 ultrasonic transducer
超声检查 ultrasonic inspection
超声脉冲 ultrasonic pulse
超声能 ultrasonic energy
超声清洗 ultrasonic cleaning; ultrasonic washing
超声速 supersonic speed
超声探伤 ultrasonic inspection
超声透镜 ultrasonic lens
超声无损检测 ultrasonic nondestructive testing
超声物理学 ultrasonic physics
超声学 ultrasonics
超声振动 ultrasonic vibration
超声致光衍射 ultrasonic light diffraction
超声钻 ultrasonic drill
超顺磁性 superparamagnetism
超速离心机 ultracentrifuge
超塑性 superplasticity
超外差接收机 superheterodyne receiver
超微晶 ultracrystallite
超文本 hypertext
超细颗粒 ultrafine grain
超弦理论 superstring theory
超显微镜 ultramicroscope
超新星 supernova
超星系 supergalaxy
超星系团 supercluster
超压 superpressure
超音速 supersonic speed
超引力 supergravity
超越方程 transcendental equation
超越数 transcendental number
超折射 superrefraction
超真空 ultravacuum
超质子 superproton
超重 overweight
超重核 superheavy nucleus
超重氢 tritium
超重元素 superheavy element; supertransuranics
超子 hyperon
潮解 deliquescence
潮汐 tide
潮汐能 tidal energy
尘核 dust core
沉积 deposition
沉降物 fallout
陈化 aging
晨昏线 terminator
衬比度 contrast
成核 nucleation
成键 bonding
成键电子 bonding electron
成像 imaging

乘数 multiplier
程 distance
程差 path difference
程长 path length
程序 program
秤 balance；scale
吃水线 water line
弛豫 relaxation
弛豫函数 relaxation function
弛豫时间 relaxation time
持续电流 persistent current
尺寸 dimension
尺寸效应 size effect
齿距 pitch
齿轮 gear
齿轮系 train of gears
斥力 repulsive force
赤道 equator
炽热 incandescence
冲 opposition
冲程 stroke
冲度 thrust
冲击摆 ballistic pendulum
冲击波 blast wave；shock wave
冲击电流 surge current
冲击电流计 ballistic galvanometer
冲击中心 centre of percussion
冲量 impulse
充电 charge
充电时间常数 charging time constant
重构 reconstruction
重正化 renormalization
重正化群 renormalization group
抽水机 water pump
抽吸 pump
抽样空间 sample space
抽运 pump
畴 domain
臭氧层 ozone layer；ozone sphere；ozonosphere
臭氧分解 ozonolysis
臭氧计 ozonometer
臭氧浓度 ozone concentration
出(射)波 outgoing wave
出射窗 exit window
出射道 outgoing channel
出射点 eyepoint
出射度 exitance
出射角 emergence angle
出射粒子 outgoing particle
初积分 first integral
初级像差 primary aberration
初始条件 initial condition
初速(度) initial velocity
初态 initial state
初子 rishon
除气 degassing
触变性 rheopexy；thixotropy
触点(压)力 contact force
触发 trigger
触发电路 flip-flop；trigger
触发二极管 trigger diode
触发灵敏度 triggering sensitivity
触发脉冲 triggering pulse；trigger pulse
触发能 triggering energy
触发器 flip-flop；trigger
触发三极管 trigger triode
氚 tritium
氚核 triton
氚化合物 tritiated compound；tritide
穿透概率 penetration probability
穿透粒子 penetrating particle
穿透能力 penetrating power
穿透深度 penetration depth
穿越辐射 transition radiation
传播 propagation
传播常量 propagation constant
传播函数 propagator
传播矢量 propagation vector
传播速度 propagation velocity
传导 conduction
传导电流 conduction current
传导率 conductivity
传递函数 transfer function
传递矩阵 transfer matrix
传递系数 transmission coefficient

传递阻抗 transfer impedance
传动比 transmission ratio
传动装置 gear
传感器 sensor; transducer
传热 heat transfer; thermal transmission
传声性 acoustical conductivity
传输 transmission
传输媒质 transmission medium
传输频带 passband; transmission band
传输损耗 transmission loss
传输误差 transmission error
传输线 transmission line
传真电报 phototelegram; telephoto
传真电报术 phototelegraphy
传真电话机 phototelephone
传质 mass transfer
串激电机 series-wound machine
串激绕法 series winding
串激绕组 series winding
串级变压器 cascade transformer
串联 connection in series; series
串联电路 series circuits
串联共振 series resonance
串联元件 series element
串列式加速器 tandem accelerator
窗 window
垂直入射 normal incidence; vertical incidence
垂直投影 vertical projection
垂直线 vertical line
纯度 purity
纯滚动 pure rolling
纯态 pure state
磁安 abampere
磁暴 magnetic storm
磁北 magnetic north
磁变 magnetic variation
磁变仪 variometer
磁波子 magnon
磁层 magnetosphere
磁层顶 magnetopause
磁常量 magnetic constant
磁场 magnetic field
磁场强度 magnetic (field) intensity; magnetic field strength
磁场热流效应 thermomagnetic effect
磁场梯度漂移 gradient B drift
磁场线 magnetic field line
磁场线圈 field coil
磁弛豫 magnetic relaxation
磁抽运 magnetic pumping
磁畴 magnetic domain
磁存储 magnetic storage
磁带 magnetic tape
磁带录音机 magnetic tape recorder; magnetophone
磁单极子 magnetic monopole
磁导计 permeameter
磁导率 (magnetic) permeability
磁电 magnetoelectricity
磁电机 magneto
磁电效应 magnetoelectric effect
磁电学 magnetoelectricity
磁顶 magnetopause
磁动势 m. m. f. (magnetomotive force)
磁对称性 magnetic symmetry
磁粉芯 dust core
磁伏 abvolt
磁感应强度 magnetic flux density; magnetic induction
磁刚度 magnetic rigidity
磁各向异性 magnetic anisotropy
磁共振 magnetic resonance
磁光效应 magneto-optical effect
磁光学 magneto-optics
磁荷 magnetic charge
磁后效 magnetic aftereffect
磁化 magnetization
磁化机 magnetizer
磁化率 (magnetic) susceptibility
磁化器 magnetizer
磁化强度 magnetization
磁化曲线 magnetization curve
磁化学 magnetochemistry
磁机工比率 magnetomechanical

ratio
磁极 （magnetic）pole
磁极片 pole piece
磁极强度 pole strength
磁记录 magnetic recording
磁介质 magnetic medium
磁晶 magnetocrystalline
磁镜 magnetic mirror
磁矩 magnetic moment
磁聚焦 magnetic focusing
磁壳 magnetic shell
磁控管 magnetron
磁控溅射 magnetron sputtering
磁力 magnetic force
磁力泵 magnetic pump
磁-力效应 magnetomechanical effect
磁量子数 magnetic quantum number
磁流体 magnetic fluid
磁流体动力学 MHD （magnetohydrodynamics）
磁路 magnetic circuit
磁面 magnetic surface
磁敏器件 magnetosensitive sensor
磁能 magnetic energy
磁能积 magnetic energy product
磁能密度 magnetic energy density
磁黏性系数 magnetic viscosity coefficient
磁黏(滞)性 magnetic viscosity
磁扭线 twistor
磁欧 abohm
磁偶极辐射 magnetic dipole radiation
磁偶极矩 magnetic dipole moment
磁偶极子 magnetic dipole
磁盘 diskette；magnetic disk
磁泡 magnetic bubble
磁泡存储器 magnetic bubble memory
磁偏角 （magnetic）declination
磁偏转 magnetic deflection
磁屏蔽 magnetic shielding
磁瓶 magnetic bottle
磁谱 magnetic spectrum
磁谱仪 magnetic spectrometer
磁气圈 magnetosphere
磁强计 gaussmeter；magnetometer
磁强记录仪 magnetograph
磁倾计 inclinometer
磁倾角 inclination；magnetic dip
磁群 magnetic group
磁热效应 magnetocaloric effect
磁韧致辐射 magnetobremsstrahlung
磁声波 magnetosonic wave
磁声效应 magnetoacoustic effect
磁声转换器件 magnetoacoustic transfer device
磁弹效应 magnetoelastic effect
磁探针 magnetic probe
磁体 magnet
磁调谐 magnetic tuning
磁调谐激光器 magnetic tuning laser
磁铁 magnet
磁通计 fluxmeter
磁通量 magnetic flux
磁通量子 fluxion
磁通流动 flux flow
磁通流阻 flux flow resistance
磁通门磁强计 flux-gate magnetometer
磁通密度 magnetic flux density
磁通蠕变 flux creep
磁通势 m. m. f. （magnetomotive force）
磁通跳变 flux jumping
磁透镜 magnetic lens
磁位 magnetic potential
磁芯 core
磁芯损耗 core loss
磁性 magnetism
磁性半导体 magnetic semiconductor

磁性材料 magnetic material
磁性常数 magnetic constant
磁性流体 magnetic fluid
磁性气体动力学 magnetogasdynamics
磁悬浮列车 maglev (train); magnetically levitated train
磁学 magnetics; magnetism
磁硬度 magnetic rigidity
磁约束 magnetic confinement
磁跃迁 magnetic transition
磁针 magnetic needle
磁振子 magnon
磁致变温效应 magnetocaloric effect
磁致电阻 magnetoresistance
磁致电阻效应 magnetoresistance effect
磁致冷 magnetic refrigeration
磁致伸缩 magnetostriction
磁致弹性 magnetoelasticity
磁致旋光 magnetic opticity; magnetic rotation
磁滞回线 (magnetic) hysteresis loop
磁滞曲线 hysteresis curve
磁滞损耗 magnetic hysteresis loss
磁滞现象 magnetic hysteresis
磁轴 magnetic axis
磁转变 magnetic transition
磁子 magneton
磁子午圈 magnetic meridian
磁阻 (magnetic) reluctance; magnetic resistance
磁阻率 reluctivity
磁阻效应 magnetoresistance effect
次波 secondary wave
次发效应 secondary effect
次级穿透粒子 penetrating progeny
次级电子 secondary electron
次级电阻 secondary resistance
次级发射 secondary emission
次级辐射 secondary radiation
次级共振 secondary resonance
次级贯穿粒子 penetrating progeny
次级光电效应 secondary photoelectric effect
次级离子质谱学 secondary ion mass spectroscopy
次级粒子 secondary partical
次级绕组 secondary winding
次级像差 secondary aberration
次级胁强 secondary stress
次级应力 secondary stress
次级运动 secondary motion
次级子波 secondary wavelet
次极大 secondary maximum
次临界的 subcritical
次声 infrasound
次声波 infrasonic wave
次声区 infrasonic region
辏力 central force
辏力场 central field
粗晶 macrocrystal
粗粒密度 coarse-grained density
粗调(节) coarse adjustment
簇 cluster
簇射 shower
催化转化器 catalytic converter
脆性材料 brittle material
萃取 extraction
萃取势 extraction potential
淬火 quenching
存储环 storage ring
锉刀 file

Dd

达朗贝尔方程　d'Alembert equation
达朗贝尔惯性力　d'Alembert inertial force
达朗贝尔原理　d'Alembert principle
达因　dyne
打印机　printer
打字电报机　teletyper；telewriter
大爆炸论　big-bang theory
大爆炸宇宙论　big-bang cosmology
大不列颠合金　Britannia metal
大潮　spring tide
大地线　geodesic（line）
大分子　macromolecule
大功率磁控管　power magnetron
大规模集成电路　LSI circuit（large-scale integrated circuit）
大卡　Calorie；kilocalorie；kilogramme calorie
大孔径透镜　high-aperture lens；wide-aperture lens
大块结晶　macrocrystal
大气　atmosphere
大气辉光　airglow
大气污染　air pollution；airborne contamination
大气压　atmosphere；atmospheric pressure；barometric pressure
大统一　grand unification
大统一理论　GUT（grand unified theory）
大型计算机　mainframe computer
大圆　great circle
代数和　algebraic sum
代数学　algebra
带标目镜　scale eyepiece
带电　electrification；electrization
带电电路　alive circuit
带电粒子　charged particle
带电粒子反应　charged particle reaction
带电粒子束聚变　charged particle beam fusion
带电体　charged body；electrified body
带负电的　electronegative
带光谱　band spectrum
带间跃迁　band-to-band transition；interband transition
带宽　bandwidth
带通滤波器　band filter
带通滤光片　band filter
带隙　band gap
带翼卫星　paddlewheel satellite
带状谱　band spectrum
带阻滤波器　band elimination filter；band suppression filter
殆周期振荡　almost periodic oscillation
戴维南定理　Thevenin theorem
戴维孙-革末实验　Davisson-Germer experiment
丹聂尔电池　Daniell cell
单摆　mathematical pendulum；simple pendulum
单胞　unit cell
单臂电桥　Wheatstone bridge
单边带调制　single-sideband modulation
单变型液晶　monotropic liquid

crystal
单层 monolayer
单程增益 single-pass gain
单纯胁变 pure strain
单纯应变 pure strain
单(磁)畴 single domain
单次曝光 single exposure
单峰激光器 unimodal laser
单缝衍射 single-slit diffraction
单工电路 simplex circuit
单极晶体管 unipolar transistor
单极轴 uniterminal axis
单极子 monopole
单结晶体管 unijunction transistor
单晶 monocrystal; single crystal
单晶生长 growth of single crystal
单晶体 monocrystal; single crystal
单模 single mode
单模变换 unimodular transformation
单模光纤 single-mode fibre
单模激光器 unimodal laser
单色波 monochromatic wave
单色辐射 monochromatic radiation
单色光 monochromatic light
单色光镜 monochromator
单色光源 monochromatic source
单色器 monochromator
单色性 monochromaticity
单色仪 monochromator
单态 singlet state
单体 monomer
单位 unit
单位磁极 unit magnetic pole
单位电荷 unit charge
单位矢量 unit vector
单位向量 unit vector
单位质量 unit mass
单向导电 unilateral conduction
单斜晶的 monoclinic
单斜晶体 monoclinic crystal
单元 cell; unit
单原子气体 monoatomic gas
单值定理 uniqueness theorem
单轴晶体 uniaxial crystal
单子 uniton
胆甾(醇)相 cholesteric phase
胆甾型液晶 cholesteric liquid crystal
弹道 ballistic curve; trajectory
弹道摆 ballistic pendulum
弹道火箭 ballistic rocket
弹道曲线 ballistic curve
弹道学 ballistics
挡热板 heat shield
氘 deuterium; heavy hydrogen
氘核 deuteron
导波体 wave director
导出单位 derived unit
导磁极 leading pole
导带 conduction band
导电介质 conducting medium
导电纤维 conductive fibre
导电性 (electric) conductivity
导率 conductance
导纳 admittance
导热析气计 katharometer
导热性 temperature conductivity; thermal conductivity
导数 derivative
导体 (electric) conductor
导线 wire
导心 guiding centre
倒电容 elastance
倒格点 reciprocal lattice point
倒格矢 reciprocal lattice vector
倒焦距 focal power
倒介电常数 elastivity
倒相 paraphase
倒相放大器 amplifier-inverter; paraphase amplifier
倒相器 (phase) inverter
倒像 inverted image
倒像棱镜 inverting prism; reversing prism
倒易点格 reciprocal lattice
倒易点阵 reciprocal lattice

倒易空间 reciprocal space
倒装片法 flip-chip
道 channel
道尔顿 dalton
道尔顿分压定律 Dalton's law of partial pressure
道尔顿原子理论 Dalton's atomic theory
德拜 debye
德拜模型 Debye model
德拜温度 Debye temperature
德拜-休克尔方程 Debye-Hückel equation
德拜-休克尔(溶液)理论 Debye-Hückel theory
德布罗意波 de Broglie wave
德布罗意波长 de Broglie wavelength
德布罗意关系 de Broglie relation
德银 German silver; nickel silver
灯泡 light bulb
灯丝 filament
等凹透镜 equiconcave lens
等比级数 geometric series
等差级数 arithmetic progression; arithmetic series
等差数列 arithmetic series
等超额中子核素 isodiaphere
等磁力线 isodynamic line
等磁倾线 isoclinic (line)
等磁线 isomagnetic line
等电点 isoelectric point
等电离点 isoelectric point
等电位面 isopotential surface
等电子的 isoelectronic
等电子群 isoelectronic group
等幅波 continuous wave
等辐透线 isophote
等厚干涉 equal thickness interference
等厚条纹 equal thickness fringes
等厚线 isopach
等加速度 constant acceleration
等离体子 plasmon
等离子体 plasma
等离子体波导 plasmaguide
等离子体层 plasmasphere
等离子体层顶 plasmapause
等离子体二极管 plasma diode
等离子体发生器 plasmatron
等离子体放电 plasma discharge
等离子体管 plasmatron
等离子体回波 plasma echo
等离子体激光器 plasma laser
等离子体加热 plasma heating
等离子体加速器 plasma accelerator
等离子体焦点 plasma focus
等离子体刻蚀 plasma etching
等离子体扩散 plasma diffusion
等离子体频率 plasma frequency
等离子体枪 plasma gun
等离子体鞘 plasma sheath
等离子体团 plasmoid
等离子体物理学 plasma physics
等离子体振荡 plasma oscillation
等离子体直线加速器 plasma linac
等偏线 isogonal line
等强度线 isophote
等倾干涉 equal inclination interference
等倾条纹 equal inclination fringes
等倾线 isoclinic (line)
等容过程 isochoric process
等容线 isochore
等熵压缩 isentropic compression
等渗压的 isotonic
等时性 isochronism
等时振荡 isochronous oscillation
等势面 equipotential surface; isopotential surface
等势线 equipotential line; isopotential line
等速 constant speed; uniform speed
等体积过程 isochoric process
等体积线 isochore
等位线 equipotential line

等温过程 isothermal process
等温线 isotherm; thermoisopleth
等相面 cophasal surface
等效电路 equivalent circuit
等效电子 equivalent electron
等效焦距 equivalent focal length
等效力系 equivalent force system
等效原理 equivalence principle; principle of equivalence
等斜线 isocline
等斜褶皱 isoclinal fold
等压线 isobar
等晕区 isoplanatic region
等张比容 parachor
等照度线 isophote
等中子异位素 isotone
等轴的 isometric
低潮 neap tide
低共熔混合物 eutectic mixture
低能电子衍射 LEED (low-energy electron diffraction)
低频 LF (low frequency)
低频扬声器 woofer
低渗溶液 hypotonic solution
低维固体 low dimensional solid
低维物理学 low dimensional physics
低温保持器 cryostat
低温抽气泵 cryogenic pump
低温等离子体 low-temperature plasma
低温电子学 cryo(e)lectronics
低温工程 cryogenic engineering
低温管 cryotron
低温恒温器 cryostat
低温计 cryometer; low-temperature thermometer
低温晶体管 cryosistor
低温物理学 low temperature physics
低温吸附 cryosorption
低温学 cryogenics
狄拉克方程 Dirac equation
笛卡儿坐标 Cartesian coordinates
底夸克 bottom quark; b-quark
底片 negative plate
底座 foot
地 earth
地磁 geomagnetism; terrestrial magnetism
地磁场 earth's magnetic field; geomagnetic field
地磁赤道 magnetic equator
地磁记录仪 magnetograph
地磁学 geomagnetism; terrestrial magnetism
地磁要素 magnetic elements
地磁仪 magnetometer
地动说 heliocentric theory
地面电波 ground wave
地面辐射 terrestrial radiation
地面望远镜 terrestrial telescope
地面折射 terrestrial refraction
地平经度 azimuth
地平纬度 altitude
地球大气 earth's atmosphere
地球反照 earthshine
地球辐射 terrestrial radiation
地球年龄 age of the earth
地球物理学 geophysics
地球中心说 geocentric theory; geocentricism
地热能 geothermal energy
地线 earth wire
地心说 geocentric theory
地心体系 geocentric system
地心引力 gravity
地震 earthquake; seism
地震波 seismic wave
地震学 seismology
地震仪 seismograph
地震中心 epicentre
地址译码器 address decoder
第二布里渊区 second Brillouin zone
第二焦点 focus in image space
第二类超导体 type Ⅱ superconductor
第二类永动机 perpetual motion

machine of the second kind
第二宇宙速度 second cosmic velocity
第三宇宙速度 third cosmic velocity
第四维 fourth dimension
第一泛音 first overtone
第一积分 first integral
第一焦点 focus in object space
第一类超导体 type Ⅰ superconductor
第一类永动机 perpetual motion machine of the first kind
第一声 first sound
第一性原理 first principle
第一宇宙速度 first cosmic velocity；orbital speed；orbital velocity
缔合律 associative law
点 point
点电荷 point charge
点对称 central symmetry；point symmetry
点火 ignition
点火能 ignition energy
点火温度 ignition temperature
点积 dot product；scalar product
点接触结 point-contact junction
点接触式晶体管 point-contact transistor
点晶格 point lattice
点粒子 point particle
点缺陷 point defect
点群 point group
点阵 (point) lattice
点阵理论 lattice theory
点阵能 lattice energy
点阵矢量 lattice vector
点阵振动 lattice vibration
点阵座 lattice site
电 electricity
电饱和 electric saturation
电报 telegram；telegraph
电报机 telegraph
电冰箱 refrigerator
电场 electric field
电场强度 electric field intensity
电场强度计 electrometer of field strength
电池 battery；(electric) cell
电池充电器 battery charger
电池铅板 grid
电池组 battery
电传 telefax
电传打字机 teletyper
电传导 electric conduction
电磁 electromagnetism
电磁半径 electromagnetic radius
电磁泵 electromagnetic pump
电磁波 electromagnetic wave
电磁波谱 electromagnetic wave spectrum
电磁场 electromagnetic field
电磁场应力张量 Maxwell stress tensor
电磁场张量 electromagnetic field tensor
电磁单位 e.m.u. (electromagnetic units)
电磁动量 electromagnetic momentum
电磁多极辐射 electromagnetic multipole radiation
电磁辐射 electromagnetic radiation
电磁辐射散射 scattering of electromagnetic radiation
电磁感应 electromagnetic induction
电磁感应加速器 rheotron
电磁理论〖光的〗 electromagnetic theory of light
电磁能量 electromagnetic energy
电磁衰变 electromagnetic decay
电磁体 electromagnet
电磁铁 electromagnet
电磁相互作用 electromagnetic interaction
电磁学 electromagnetics；electromagnetism

电磁质量 electromagnetic mass
电磁阻尼 electromagnetic damping
电导 conductance
电导率 (electric) conductivity
电导体 electric conductor
电导系数 conductance
电灯 electric lamp
电动(次)序 electromotive series
电动发电机 motor generator
电动机 (electric) motor
电动力学 electrodynamics
电动势 e.m.f.(electromotive force)
电动势序列 electromotive series
电动序 electromotive series
电度表 watthour meter
电镀(术) electroplating
电多极矩 electric multipole moment
电多极子 electric multipole
电风 electric wind
电感 inductance
电感耦合 inductive coupling
电感器 inductor
电感线圈 induction coil
电工学 electrotechnics
电功率 electric power
电光晶体 electrooptical crystal
电光偏转 electrooptical deflection
电光调制 electrooptical modulation
电光系数 electrooptical coefficient
电光效应 electrooptical effect
电光学 electrooptics
电荷 (electric) charge
电荷半径 charge radius
电荷补偿 charge compensation
电荷不变性 charge invariance
电荷不对称性 charge asymmetry
电荷对称性 charge symmetry
电荷共轭 charge conjugation
电荷交换 charge exchange
电荷密度 charge density
电荷密度波 CDW(charge density wave)
电荷耦合器件 CCD(charge coupled device)
电荷耦合装置 CCD(charge coupled device)
电荷守恒 conservation of charge
电荷守恒定律 law of conservation of charge
电荷算符 charge operator
电荷无关性 charge independence
电荷宇称不守恒 CP violation
电荷宇称守恒 CP invariance
电荷注入 charge injection
电荷注入器件 CID(charge injection device)
电荷转移 charge transfer
电荷转移过程 charge transfer process
电荷转移器件 charge transfer device
电弧 electric arc
电弧放电 arc discharge
电弧炉 electric-arc furnace
电花插头 spark plug
电化当量 electrochemical equivalent
电化电池 electrochemical cell
电化序 electrochemical series; electromotive series
电化学 electrochemistry
电化学发光 electrochemiluminescence
电话 telephone
电火花 electric spark
电击致死 electrocution
电极 electrode; pole
电极化 electric polarization
电极化率 electric susceptibility
电极化强度 electric polarization
电极强度 pole strength
电极效应 pole effect
电键 switch
电解 electrolysis
电解池 electrolytic cell

电解电容 electrolytic capacitor
电解淀积物 electrodeposit
电解分离 electrolytic separation
电解腐蚀 electrolytic corrosion
电解精炼 electrolytic refining
电解精制 electrolytic refining
电解气 electrolytic gas
电解溶液 electrolyte
电解式整流器 electrolytic rectifier
电解箱 electrolytic cell
电解液 electrolyte
电解质 electrolyte
电介质 dielectric
电介质极化 dielectric polarization
电介质加热 dielectric heating
电聚 voltolization
电抗 reactance
电缆 cable
电离 ionization
电离比值 specific ionization
电离层 ionosphere
电离层波 ionospheric wave
电离电流 ionization current
电离电势 IP（ionization potential）
电离电位 IP（ionization potential）
电离辐射 ionizing radiation
电离能 ionization energy
电离谱学 ionization spectroscopy
电离室 ionization chamber
电离箱 ionization chamber
电离压力计 ionization gauge
电离压强计 ionization gauge
电离真空规 ionization gauge
电力 electric force；electric power
电力传输 power transmission
电力传输线 power line
电力输送 electric power transmission
电力线 electric line of force
电量 electric quantity
电量计 coulometer；voltameter
电铃 electric bell
电流 （electric）current
电流表 ammeter；ampere-meter
电流穿透深度 penetration of current
电流计 galvanometer
电流密度 current density
电流强度 current intensity
电流体力学 EHD（electrohydrodynamics）
电流天平 current balance
电流线 electric streamline
电流元 current element
电路 （electric）circuit
电路图 circuit diagram
电纳 susceptance
电能 electric energy
电偶极辐射 electric dipole radiation
电偶极跃迁 electric dipole transition
电偶极子 electric dipole
电偏转 electric deflection
电平 level
电平密度 level density
电气安全 electric safety
电气化 electrification；electrization
电气绝缘 electric insulation
电器 electric appliance
电桥 electric bridge
电热效应 electrocaloric effect
电热学 electrothermics
电容 capacitance；capacity
电容率 dielectric constant；permittivity
电容器 capacitor；condenser
电弱统一理论 electro-weak unified theory
电弱相互作用 electro-weak interaction
电色层分离法 electrochromatography
电色谱法 electrochromatography

电渗透 electroosmosis
电渗析 electrodialysis
电渗现象 electroosmosis
电生物学 electrobiology
电声倒易定理 electroacoustical reciprocity theorem
电声互易原理 electroacoustical reciprocity theorem
电声器件 acoustoelectronic device；electroacoustic device
电声系统 electroacoustic system
电声学 electroacoustics
电势 electric potential
电势差 electric potential difference
电势差计 potentiometer
电势分配器 potential divider
电势降 potential drop
电视(机) teleset；televisor；TV (television)
电视摄像机 television camera
电视卫星 telestar
电视显像管 television tube
电视银幕 telescreen
电枢 armature
电刷 brush
电刷放电 brush discharge
电四极辐射 electric quadrupole radiation
电四极矩 electric quadrupole moment
电四极子 electric quadrupole
电弹常量 electroelastic constant
电通量 electric flux
电通量密度 electric flux density
电透法 cataphoresis
电位 electric potential
电位差 electric potential difference
电位差计 potentiometer
电位器 potential divider；potentiometer
电位序 electromotive series
电位移 electric displacement
电位移通量 electric flux
电线 wire
电像法 method of images
电信 telecommunication
电性流 charged current
电学 electricity
电压 voltage
电压比较器 voltage comparator
电压表 voltmeter
电压放大器 voltage amplifier
电压分配器 voltage divider
电冶金学 electrometallurgy
电泳 cataphoresis；electrophoresis
电涌阻抗 surge impedance
电源 power source；power supply
电源组 power pack
电跃迁 electric transition
电晕 corona
电晕放电 corona discharge
电致发光 electroluminescence
电致伸缩 electrostriction
电致双折射 electric birefringence
电中性 electric neutrality；electroneutrality
电铸 electroforming
电子 electron；negaton
电子包 packet of electrons
电子倍增器 electron multiplier
电子弹道学 electroballistics
电子电压表 electronic voltmeter
电子定时器 electrotimer
电子伏(特) eV (electronvolt)
电子伏特计 electronic voltmeter
电子俘获 electron capture
电子感应加速器 betatron；rheotron
电子管 electron tube；(vacuum) tube；(vacuum) valve
电子管参数 tube parameter
电子管放大器 tube amplifier
电子管振荡发生器 tube generator
电子管振荡器 tube oscillator
电子光学 electron optics
电子回旋加速器 microtron

电子加速器 electron accelerator
电子开关 electronic switch
电子控制 electronic control
电子能耗谱学 electron energy loss spectroscopy
电子能耗谱仪 EELS (electron energy loss spectrometer)
电子能量分析器 electron energy analyzer
电子偶素 positronium
电子气 electron gas
电子枪 electron gun
电子亲和势 electron affinity
电子亲和性 electron affinity
电子群 packet of electrons
电子摄像机 electron camera
电子–声子互作用 electron-phonon interaction
电子受激脱附 electron stimulated desorption
电子束 electron beam
电子顺磁共振 EPR (electron paramagnetic resonance)
电子探针微量分析 EPM (electron probe microanalysis)
电子探针微区分析 EPM (electron probe microanalysis)
电子同步加速器 electron synchrotron
电子透镜 electron lens
电子望远镜 electron telescope
电子显微镜 electron microscope
电子显微照片 electron micrograph
电子相变 electron phase transition
电子学 electronics
电子雪崩电离 electron avalanche ionization
电子衍射 electron diffraction
电子衍射技术 electron diffraction technique
电子衍射仪 electron diffractometer
电子异构体 electromer
电子云 electron cloud
电子噪声 electronic noise
电子振荡 electron oscillation
电子自旋 electron spin
电子自旋共振 ESR (electron-spin resonance)
电阻 (electric) resistance; resistor
电阻表 ohmmeter
电阻高温计 resistance pyrometer
电阻加热 ohmic heating
电阻率 resistivity; specific resistance
电阻器 resistor
电阻损耗 ohmic loss
电阻温度计 resistance thermometer
电阻系数 specific resistance
电阻箱 resistance box
垫圈 packing ring
淀积 deposition; precipitation; sedimentation
迭代法 iteration method
迭代过程 iterative process
叠加原理 principle of superposition; superposition principle
叠片铁芯 laminated core
叠栅条纹 moiré fringe
叠像测距仪 coincidence range finder
叠像棱镜 coincidence prism
顶点 zenith
顶角 vertex angle
定标电路 scaling circuit
定标器 scaler
定常流(动) steady flow
定常态 steady state
定常约束 scleronomic constraint; steady constraint
定点 fixed point
定点转动 rotation around a fixed point
定滑轮 fixed pulley
定积分 definite integral
定理 theorem

定律 law; rule
定率计 rate meter
定片 stator
定日镜 heliostat
定时电路 timing circuit
定时脉冲发生器 timing pulse generator
定时器 timer
定态 stationary state
定态波 stationary wave
定态场 stationary field
定态系综 stationary ensemble
定态薛定谔方程 stationary Schrödinger equation
定态宇宙理论 steady-state theory of the universe
定体(积)比热 specific heat at constant volume
定天镜 coelostat
定位 localization
定相 phasing
定向光束 directed-beam
定向天线 directional aerial; directive aerial
定压比热 specific heat at constant pressure
定域化 localization
定域理论 local theory
定则 rule
定轴转动 fixed-axis rotation
定子 stator
丢失质量 missing mass
动电学 electrokinetics
动滑轮 movable pulley
动理方程 kinetic equation
动理模 kinetic mode
动理势 kinetic potential
动理学 kinetics
动理学方程 kinetic equation
动力单元 power pack
动力反应堆 power reactor
动力黏度 kinetic viscosity
动力平衡 dynamic equilibrium
动力系 dynamical system
动力学 dynamics; kinetics
动力学变量 dynamical variable
动力学对称 dynamical symmetry
动力学对称破缺 dynamical symmetry breaking
动力学方程 kinetic equation
动力学矩阵 dynamical matrix
动力学系统 dynamical system
动力站 power installation
动力装置 power installation
动量 momentum
动量表象 momentum representation
动量矩 angular momentum; moment of momentum
动量矩守恒定律 angular-momentum conservation law; law of conservation of moment of momentum
动量空间 momentum space
动量守恒 conservation of momentum; momentum conservation
动量守恒定律 law of conservation of momentum
动摩擦 kinetic friction
动能 kinetic energy
动圈式仪表 moving-coil instrument
动生电动势 motional electromotive force
动态范围 dynamic range
动态平衡 dynamic equilibrium; mobile equilibrium
动态散射 dynamic scattering
动态系统 dynamical system
动铁式仪表 moving-iron instrument
动性 mobility
动压(力) dynamical pressure
动压强 dynamical pressure
动作电位 action potential
独立粒子模型 independent-particle model
独立振幅 independent amplitude
杜隆-珀蒂定律 Dulong-Petit law
杜瓦瓶 Dewar flask; vacuum

flask
度 degree；kWh（kilowatt-hour）
度规张量 metric tensor
度量衡学 metrology
度量衡制 metrology
渡越辐射 transition radiation
渡越损失 transition loss
镀层 cladding
端电压 terminal voltage
端子 post；terminal
短程力 short-range force
短路 short circuit
短缺质量 missing mass
断层 dislocation
断点 scission point
断开 turn-off
断裂强度 breaking strength；fracture strength
断路 open circuit
断路开关 circuit breaker
断路器 （circuit）breaker
锻铁 wrought iron
堆垛层错 stacking fault
堆集密度 packing density
对产生 pair creation；pair production
对称 symmetry
对称定律 law of symmetry
对称面 plane of symmetry
对称性 symmetry
对称轴 axis of symmetry
对地静止轨道 geostationary orbit
对分布函数 pair distribution function
对流 advection；convection
对流层 troposphere
对流电流 convection current
对偶场 dual field
对偶空间 dual space
对偶性 duality
对偶张量 dual tensor
对日照 Gegenschein
对数 logarithm
对数比例 logarithmic scale
对数标尺 logarithmic scale
对数标度 logarithmic scale
对数级数 logarithmic series
对头碰撞 central collision；head-on collision
对易关系 commutation relation
对易式 commutator
对应态 corresponding state
对应原理 correspondence principle
对映形态 enantiomorphism
对映异构体 enantiomers
对撞机 collider
对撞束 colliding beam
吨 ton
钝化 deactivation；passivation
多边形 polygon
多层超薄共格结构 LUCS（layered ultrathin coherent structure）
多层介电膜 multilayer dielectric film
多层吸附 multilayer adsorption
多掺激光器 alphabet laser
多重散射 multiple scattering
多重数 multiplicity
多重态 multiplet
多重线 multiplet
多重性 multiplicity
多畴 polydomain
多道分光计 multichannel spectrometer
多道分析器 multichannel analyzer
多道谱仪 multichannel spectrometer
多电极 polyelectrode
多电子 polyelectron
多方过程 polytropic process
多方指数 polytropic exponent
多光束 multiple beam
多光柱 multiple beam
多光子 multiphoton
多级火箭 multistage rocket
多级激光器 cascade laser
多极辐射 multipole radiation
多极跃迁 multipole transition

多角形 polygon
多晶 polycrystal
多晶硅 polysilicon
多晶体 polycrystal
多路存取系统 multiaccess system
多媒体 multimedia
多面角 polyhedral angle；solid angle
多面体 polyhedron
多模光纤 multimode optical fibre
多模激光器 multimode laser
多模腔 multimode cavity
多频振荡器 multivibrator
多普勒频移 Doppler shift
多普勒频移衰减法 Doppler shift attenuation method
多普勒谱线增宽 Doppler broadening
多普勒速度计 Doppler velocimeter
多普勒效应 Doppler effect
多普勒增宽 Doppler broadening
多色差透镜 hyperchromatic lens
多色辐射 polychromatic radiation
多色光 polychromatic light
多数规则 majority rule
多数载流子 majority carrier
多体问题 many-body problem
多向色性 pleochroism
多项的 polynomial
多项式 polynomial
多相 multiphase
多芯复合超导体 multifilamentary composite superconductor
多用电表 multimeter；universal meter
多子 majority carrier
多子浓度 majority carrier density

Ee

俄歇电子　Auger electron
俄歇效应　Auger effect
额定电压　rated voltage
额定功率　nominal power
额定值　nominal value
厄米-高斯波型　Hermite-Gauss mode
厄米-高斯光束　Hermite-Gauss beam
厄米-高斯模　Hermite-Gauss mode
厄米共轭　Hermitian conjugate
厄米算符　Hermitian operator
扼流圈　choke
扼流线圈　choking coil
尔格　erg
二冲程发动机　two-cycle engine
二重性　duality
二次方程　quadratic equation
二次曲线　conic
二次效应　secondary effect
二次谐波　second harmonic
二分点　equinoxes
二分量中微子理论　two-component neutrino theory
二分裂变　binary fission
二级同位素　secondary isotope
二级相变　second-order phase transition
二极的　ambipolar
二极管　diode
二极管激光器　diode laser
二进制记数法　binary notation
二进制全息图　binary hologram
二进制数字　binary digit
二面体　dihedron
二色性　dichroism; dichromatism
二十面体　icosahedron
二十面体群　icosahedral group
二体碰撞　binary collision
二体问题　two-body problem
二维光栅　two-dimensional grating
二向色镜　dichroic mirror
二向色性　dichroism
二项式定理　binomial theorem
二项展开式　binomial expansion
二象性　duality
二氧化硅　silicon dioxide
二氧化碳　carbon dioxide
二氧化碳激光器　carbon dioxide laser
二元全息图　binary hologram
二元性　duality
二至点　solstices

Ff

发报机 transmitter
发电厂 power plant; power station
发电机 generator
发电站 power plant; power station
发动机 engine; motor
发光 luminescence
发光度 luminosity
发光二极管 LED (light-emitting diode)
发光强度 luminous intensity
发光摄谱仪 optical emission spectrograph
发光体 luminophor; twinkler
发光星体 photostar
发光学 luminescence
发散 div (divergence)
发散波 divergent wave
发散级数 divergent series
发散角 angle of divergence
发散镜 diverging mirror
发散透镜 diverging lens
发射 emission; transmission
发射本领 emissive power
发射度 emittance
发射跟随器 emitter follower
发射功率 emissive power
发射光谱 emission spectrum
发射机 transmitter
发射极 emitter
发射极耦合逻辑 ECL (emitter-coupled logic)
发射率 emissivity
发射谱 emission spectrum
发射器 emitter
发射强度 emissive power
发射天线 transmitting aerial
发生器 generator
阀 valve
法包线 evolute
法布里-珀罗标准具 Fabry-Pérot etalon
法布里-珀罗标准器 Fabry-Pérot etalon
法布里-珀罗干涉仪 Fabry-Pérot interferometer
法布里-珀罗滤波器 Fabry-Pérot filter
法拉 F (farad)
法拉第常数 Faraday constant
法拉第磁秤 Faraday balance
法拉第电磁感应定律 Faraday's laws of electromagnetic induction
法拉第定律 Faraday's laws
法拉第恒量 Faraday constant
法拉第笼 Faraday cage
法拉第扭秤 Faraday balance
法拉第屏蔽罩 Faraday cage
法拉第(涡流)效应 Faraday effect
法拉第旋转 Faraday rotation
法拉第圆筒 Faraday cylinder
法线 normal; vertical line
法向加速度 normal acceleration
法向力 normal force
法向入射 vertical incidence
法向应力 normal stress
砝码 poise
帆板卫星 paddlewheel satellite
反 K 介子 antikaon
反玻色子 antiboson
反差 contrast

反差度 image definition
反常 anomaly
反常磁矩 abnormal magnetic moment
反常扩散 anomalous diffusion
反常流 abnormal current
反常塞曼效应 anomalous Zeeman effect
反常散射 anomalous scattering
反常色散 anomalous dispersion
反常输运 anomalous transport
反常吸收 abnormal absorption; anomalous absorption
反冲 recoil
反冲电子 recoil electron
反磁性学 diamagnetism
反氘核 antideuteron
反电动势 back e. m. f.; counter e. m. f.; opposing e. m. f.
反动力 reaction
反对称 skew symmetry
反对数 antilogarithm
反费米子 antifermion
反光镜 optical reflector
反函数 inverse function
反核 antinucleus
反夸克 antiquark
反馈 feedback
反馈电路 backfeed loop
反扩散 antidiffusion
反粒子 antiparticle
反平行向量 antiparallel vectors
反平行自旋 antiparallel spins
反轻子 antilepton
反射 reflection; reverberation
反射比 reflectance
反射波 reflected wave
反射成像 catoptric imaging
反射定律 law of reflection; reflection law
反射光栅 reflection grating
反射光学 catoptrics
反射角 angle of reflection; reflection angle
反射镜 mirror; reflector; reverberator
反射炉 reverberatory furnace
反射率 reflectivity
反射器 reflector; reverberator
反射望远镜 reflecting telescope; reflector
反射系数 reflectance; reflection coefficient; reflectivity
反射线 reflected ray
反射性 reflectivity
反射折射光学系统 catadioptric system
反射折射望远镜 catadioptric telescope
反世界 antiworld
反弹 recoil
反铁磁性 antiferromagnetism
反铁电体 antiferroelectrics
反推力 reverse thrust
反物质 antimatter
反物质世界 antiworld
反向二极管 backward diode
反向扩散 back diffusion
反向自旋 opposite spins
反向作用力 opposite acting force
反相 antiphase; opposite phase
反相器 (phase) inverter
反相位 antiphase
反像 mirror image; reversed image
反像棱镜 reversing prism
反星系 antigalaxy
反型层 inversion layer
反演 inversion
反应 reaction
反应 Q 值 reaction Q-value
反应堆 reactor
反应能 reaction energy
反应热 heat of reaction
反宇称性 opposite parity
反原子 antiatom
反照率 albedo
反折系统 catadioptric system
反质子 antiproton
反中微子 antineutrino
反中子 antineutron

反重子 antibaryon
反转点 turning point
反转温度 inversion temperature
反作用 reaction
反作用力 reaction force
反作用推进 reaction propulsion
反作用原理 reaction principle
泛音 overtone；partial
范艾伦带 radiation belt；Van Allen belt
范德格拉夫加速器 Van de Graaff accelerator
范德格拉夫起电机 Van de Graaff generator
范德瓦耳斯方程 van der Waals' equation
范德瓦耳斯力 van der Waals' force
范性 plasticity
范性形变 plastic deformation
方 phon
方差 variance
方差分析 analysis of variance
方解石 calcite
方均根 RMS（root mean square）
方均根电流 RMS current
方均根电压 RMS voltage
方均根速率 RMS speed
方均根误差 RMSE（root mean square error）
方均根值 RMS value
方块棱镜 block prism
方位 azimuth
方位角 azimuth
方位偏振仪 azimuth polarimeter
方形波 square wave
防静电材料 antistatic material
防静电器 parastat
防事故外壳 containment shell
防震台 isolation table
防撞压损区 crumple zone
仿星器 stellarator
仿真 simulation
放大 magnification
放大镜 magnifier；magnifying glass
放大率 amplification ratio；magnification；magnifying power
放大率色差 chromatism of magnification
放大器 amplifier
放大系数 amplification coefficient
放大自发发射 amplified spontaneous emission
放电 discharge
放电电阻 discharge resistance
放电管 discharge tube
放能的 exoergic
放能反应 exothermic reaction
放热性 exothermicity
放射 emanation；radiation
放射产生的 radiogenic
放射化学 radiochemistry
放射量测定器 dosimeter
放射生物学 radiobiology
放射示踪法 radioactive-tracer method
放射衰变 radioactive decay
放射系 radioactive series
放射性 radiativity；radioactivity
放射性沉降物 radioactive fallout
放射性的 radioactive
放射性废物 radioactive waste
放射性核 active nucleus
放射性核素 radioactive nuclide；radionuclide
放射性活度 activity；radioactivity
放射性鉴年法 radioactive dating
放射性年龄 radioactive age
放射性平衡 radioactive equilibrium
放射性散落物 radioactive fallout
放射性示踪物 radioactive tracer
放射性碳定年 radiocarbon dating
放射性碳计时 radiocarbon

dating
放射性同位素 radioactive isotope; radioisotope
放射性指示剂 radioactive tracer
放射学 radiology
放射源 radioactive source
放射治疗 radiotherapy
放射自显影法 autoradiography
飞机 plane
非保守力 non-conservative force
非本征半导体 extrinsic semiconductor
非本征铁电体 improper ferroelectrics
非本征吸收 extrinsic absorption
非常波 extraordinary wave
非常光 extraordinary light
非常光线 extraordinary ray
非常射线 extraordinary ray
非常折射率 extraordinary refractive index
非电路 NOT circuit
非定常约束 rheonomic constraint; unsteady constraint
非定态 nonstationary state
非定域系 nonlocalized system
非对心碰撞 oblique impact
非对映异构物 diastereoisomers
非惯性系 noninertial system
非化学计量化合物 nonstoichiometric compound
非极化的 unpolarized
非简谐性 anharmonicity
非金属 nonmetal
非晶的 amorphous
非晶介电体 amorphous dielectrics
非晶态 amorphous state
非晶态固体 amorphous solid
非晶体 amorphous matter; noncrystal
非静态过程 non-static process
非聚变中子 false neutron
非绝热的 nonadiabatic
非理想气体 imperfect gas
非立方(系)的 noncubic
非门 NOT gate
非牛顿流体 non-Newtonian fluid
非欧几里得几何学 non-Euclidean geometry
非偏振的 unpolarized
非偏振光 nonpolarized light; unpolarized light
非平衡热力学 nonequilibrium thermodynamics
非平衡态 nonequilibrium state
非平衡统计力学 nonequilibrium statistical mechanics
非平衡载流子 nonequilibrium carrier
非奇异矩阵 nonsingular matrix
非球面透镜 non-spherical lens
非热辐射 nonthermal radiation
非热核中子 false neutron
非束缚态 unbound state
非弹性(度) inelasticity
非弹性碰撞 inelastic collision
非弹性散射 inelastic scattering
非弹性散射截面 inelastic scattering cross-section
非弹性系数 inelasticity
非完全熔合 incomplete fusion
非完全弹性碰撞 imperfect elastic collision
非稳态 non-steady state; unsteady state
非线性光学 nonlinear optics
非线性响应 nonlinear response
非线性效应 nonlinear effect
非相干成像 incoherent imaging
非相干辐射 noncoherent radiation
非相干光 incoherent light
非相干散射 incoherent scattering
非相干性 incoherence
非相干照明 incoherent illumination
非谐力 anharmonic force
非谐性 anharmonicity
非谐振荡器 anharmonic

oscillator
非谐振动 anharmonic vibration
非谐振子 anharmonic oscillator
非有心力 non-central force
非匀速运动 non-uniform motion
非周期性 aperiodicity
菲 fresnel
菲茨杰拉德收缩 Fitzgerald contraction
菲克定律 Fick law
菲涅耳 fresnel
菲涅耳公式 Fresnel formula
菲涅耳双镜 Fresnel bimirror
菲涅耳双棱镜 Fresnel biprism
菲涅耳透镜 Fresnel lens
菲涅耳衍射 Fresnel diffraction
菲佐法 Fizeau's method
菲佐实验 Fizeau experiment
菲佐条纹 Fizeau fringe
斐波那契数 Fibonacci numbers
沸点 b. p.（boiling point）
沸点升高 elevation of boiling point
沸点升高常数 ebullioscopic constant
沸点图 boiling point diagram
沸水反应堆 BWR（boiling-water reactor）
沸腾 boiling
费恩曼图 Feynman diagram
费马原理 Fermat('s) principle
费米 fermi
费米常数 Fermi constant
费米-狄拉克积分 Fermi-Dirac integral
费米-狄拉克统计法 Fermi-Dirac statistics
费米动量 Fermi momentum
费米函数 Fermi function
费米面 Fermi surface
费米能级 Fermi level
费米球 Fermi sphere
费米相互作用 Fermi interaction
费米子 fermion
费米子弦 fermionic string
镄 Fm（fermium）
分贝(耳） dB（decibel）
分辨本领 resolving power
分辨率 resolution；resolving power
分辨时间 resolving time
分波法 method of partial waves
分布力 distributed force
分布量 abundance
分步透镜 stepped lens
分层介质 stratified medium
分点岁差 precession of the equinoxes
分度镜 graticule
分光光度计 spectrophotometer
分光计 spectrometer
分光镜 dichroic mirror；spectroscope
分光双星 spectroscopic binary
分划板 graticule
分级电位计 step potentiometer
分解 resolution
分界线 separatrix
分离过程 detachment process
分离能 separation energy
分力 component force
分立本征值 discrete eigenvalue
分立谱 discrete spectrum
分量 component
分裂 fission
分流 effusion；shunt；split stream
分流器 shunt
分路 shunt
分配律 distributive law
分配器 distributor
分气 atmolysis
分散力 dispersion force
分散相 dispersed phase
分色镜 dichroic mirror
分色性 dichroism
分时 time sharing
分矢量 component vectors
分速度 component velocity
分维 fractal dimension
分析镜 analyzer
分析力学 analytical mechanics

分析器 analyzer
分析天平 analytical balance
分像棱镜 separating prism
分形 fractal
分形测度 fractal measure
分形体 fractal
分形维数 fractal dimension
分压力 partial pressure
分压器 potential divider; voltage divider
分压强 partial pressure
分振幅法 division of amplitude
分至年 tropical year
分钟 minute
分子 molecule
分子场理论 molecular field theory
分子磁体 molecular magnet
分子电流 molecular current
分子电子学 molecular electronics
分子动理论 kinetic theory
分子固体 molecular solid
分子光谱 molecular spectrum
分子光谱学 molecular spectroscopy
分子轨函数 molecular orbital
分子假说 molecular hypothesis
分子间力 intermolecular force
分子结构 molecular structure
分子晶体 molecular crystal
分子理论 molecular theory
分子量 molecular weight; relative molecular mass
分子流 molecular flow
分子散射 molecular scattering
分子筛 molecular sieve
分子式 molecular formula
分子束 molecular beam
分子束外延 MBE (molecular beam epitaxy)
分子说 molecular theory
分子体积 molecular volume
分子物理学 molecular physics
分子运动论 kinetic theory
分子蒸馏 molecular distillation
粉末冶金学 powder metallurgy
粉芯 dust core
丰度 abundance
风力 wind power
风力发电机 aerogenerator
风力计 anemometer; wind gauge
风能 wind energy
风速计 anemometer; wind gauge
风闸 airbrake
峰高 peak height
峰化器 peaker
峰能 peak energy
峰值 amplitude; peak (value)
峰值带 peak strip
峰值高度 peak height
峰值功率 peak power
峰值脉冲幅度 peak pulse amplitude
峰值能量 peak energy
峰值限制器 peak limiter
峰值响应 peak response
峰值因数 peak factor
峰值载荷 peak load
夫琅禾费谱线 Fraunhofer line
夫琅禾费全息图 Fraunhofer hologram
夫琅禾费衍射 Fraunhofer diffraction
敷霜 blooming
弗仑克尔缺陷 Frenkel defect
伏 V (volt)
伏安 volt-ampere
伏安表 volometer
伏安法 voltmeter-ammeter method
伏安特性曲线 volt-ampere characteristics
伏打电 voltaic electricity
伏打电池 voltaic battery; voltaic cell
伏打电堆 voltaic pile
伏特 V (volt)
伏特安培 volt-ampere
伏特计 voltmeter

伏特数 voltage
俘获 capture；trap
俘获截面 capture cross-section
俘获离子 trapped ion
俘获粒子 trapped particle
俘获粒子不稳定性 trapped particle instability
浮秤 hydrometer
浮动 floating
浮环 floating ring
浮力 buoyancy（force）；floatage
浮力秤 flotation balance
浮质 aerosol
符号规约 sign convention
符合电路 coincidence circuit
符合法 coincidence method
符合光谱 coincidence spectrum
符合计数 coincidence counting
符合结构 coincidence structure
符合率 coincidence rate
幅度检波 amplitude detection
幅光栅 amplitude grating
辐角 argument
辐解作用 radiolysis
辐射 emanation；radiation
辐射安全 radiation safety
辐射不透明度 radio-opacity
辐射长度 radiation length
辐射出射度 radiant exitance
辐射带 radiation belt；Van Allen belt
辐射的量子理论 quantum theory of radiation
辐射电阻 radiation resistance
辐射发射度 radiant emittance
辐射方向图 radiation pattern
辐射防护 radiation protection
辐射俘获 radiative capture
辐射改性 radiation modification
辐射功率 radiant power
辐射剂量 radiation dose
辐射亮度 radiance
辐射量 radiant exposure
辐射率 radiance
辐射能 radiant energy
辐射偶极子 active dipole
辐射碰撞 radiative collision
辐射屏蔽 radiation shield
辐射强度 radiant intensity
辐射热 radiant heat
辐射热计 bolometer
辐射损伤 radiation damage
辐射体 radiator
辐射通量 radiant flux
辐射通量计 fluxmeter
辐射通量密度 radiant flux density
辐射透射率 radiant transmittance
辐射图 radiation pattern
辐射危害 radiation hazard
辐射微尘 radioactive fallout
辐射温度 radiation temperature
辐射吸收 radiative absorption
辐射系数 radiation coefficient
辐射性 radiativity
辐射学 radiology
辐射压(强) radiation pressure
辐射源 radiation source
辐射(致)发光 radioluminescence
辐射阻抗 radiation impedance
辐透 phot
辐照 irradiation
辐照度 irradiance
辐照损伤 irradiation damage
辐照效应 irradiation effect
福丁气压计 Fortin barometer
福克尔-普朗克方程 Fokker-Planck equation
腐蚀 corrosion
腐蚀性的 caustic
负电的 electronegative
负电荷 negative charge
负电性 electronegativity
负电子 negatron
负反馈 NF（negative feedback）；reverse feedback
负畸变 barrel distortion
负极 cathode；negative pole
负极板 negative plate
负极性 negative polarity
负介子 negative meson

负晶体 negative crystal
负离子 anion; negative ion
负脉冲 negative pulse
负能态 negative energy state
负偏差 negative deviation
负熵 negentropy
负透镜 negative lens
负位错 negative dislocation
负效放电 back discharge
负序阻抗 negative sequence impedance
负宇称 odd parity
负载 load
负质量 negative mass
负质子 negative proton
负子 negaton
负阻二极管 negative resistance diode
附加损失 parasitic loss
附着 adhesion; trap
附着力 adhesion
复摆 compound pendulum; physical pendulum
复冰 regelation
复共轭 complex conjugate
复光栅 multiple grating
复合 recombination
复合核 compound nucleus
复合核理论 compound-nucleus theory
复合核衰变 compound-nucleus decay
复合粒子 composite particle
复合滤波器 complex filter
复合受激态 exciplex
复合透镜 compound lens
复合系统 compound system
复辉 recalescence
复曲面透镜 toric lens
复数 complex number
复显微镜 compound microscope
复相干度 complex degree of coherence
复消色差透镜 apochromat; apochromatic lens
复眼透镜 fly's-eye lens
复折射率 complex index of refraction
复振幅 complex amplitude
复制光栅 replica grating
复阻抗 complex impedance
副标准 secondary standard
副光轴 secondary optical axis
副虹 secondary rainbow
副镜 secondary mirror
副绕组 secondary winding
副色 secondary colour
傅科摆 Foucault pendulum
傅里叶变换 Fourier transformation
傅里叶定律 Fourier law
傅里叶分析 Fourier analysis
傅里叶光学 Fourier optics
傅里叶级数 Fourier series
傅里叶综合(法) Fourier synthesis
覆层 cladding

Gg

伽伐尼电池 galvanic cell
伽利略变换 Galilean transformations
伽利略不变性 Galilean invariance
伽利略望远镜 Galilean telescope
伽利略相对性原理 Galilean principle of relativity
改进型气冷反应堆 AGR（advanced gas-cooled reactor）
盖革计数器 Geiger counter
盖革-米勒计数器 Geiger-Müller counter
盖-吕萨克定律 Gay-Lussac's law
盖斯勒管 Geissler tube
概率 probability
概率函数 probability function
概率积分 probability integral
概率流 probability current
概率密度 probability density
概然速度 probable velocity
概然误差 probable error
干冰 dry ice
干电池 dry cell
干胶片 wafer
干扰 interference
干扰共振 parasitic resonance
干扰计 interferometer
干涉 interference
干涉波 interference wave
干涉测量术 interferometry
干涉级 order of interference
干涉量度学 interferometry
干涉滤波器 interference filter
干涉滤光片 interference filter
干涉色 interference colour
干涉条纹 interference fringe
干涉图样 interference pattern
干涉显微镜 interference microscope
干涉项 interference term
干涉序 order of interference
干涉仪 interferometer
干湿计 psychrometer
干湿球湿度计 psychrometer；wet-and-dry bulb hygrometer
干燥器 dehumidifier
杆 lever
杆状物 shaft
坩埚 crucible
感光计 sensitometer
感光灵敏度 photosensitivity
感光性 photosensitivity
感抗 inductive reactance
感生电场 induced electric field
感生电动势 induced e. m. f.
感生电荷 induced charge
感生电流 induced current
感生发射 induced emission；stimulated emission
感生辐射 stimulated radiation
感生巨磁矩 induced giant moment
感生力矩 induced moment
感应 induction
感应场 induction field
感应电场 induced electric field
感应电动机 induction motor
感应电动势 induced e. m. f.
感应电荷 induced charge
感应电流 induced current
感应加热 induction heating
感应率 inductivity
感应器 inductor

感应圈 induction coil
感应体 inductor
感应系数 inductance
感应现象 induction
感应线圈 induction coil
刚度 rigidity；stiffness
刚度系数 rigidity；stiffness coefficient
刚体 rigid body
刚性杆 rigid rod
刚性结构 rigid structure
刚性模量 rigidity modulus
钢 steel
杠杆 lever
杠杆臂 lever arm
杠杆定则 lever rule
杠杆式天平 beam balance
高保真度 hi-fi (high fidelity)
高层大气 upper atmosphere
高差计 cathetometer
高导磁合金 Mumetal；permalloy
高电压 HT (high tension)
高电压处理 voltolization
高度 altitude
高度计 altimeter；cathetometer
高反射膜 high-reflecting film
高分辨光谱学 high-resolution spectroscopy
高分子 macromolecule
高分子物理学 polymer physics
高空气栓病 aeroembolism
高空气象学 aerology
高空生物学 aerobiology
高能电子衍射 HEED (high-energy electron diffraction)
高能核反应 high-energy nuclear reaction
高能核物理学 high-energy nuclear physics
高能天体物理学 high-energy astrophysics
高能物理学 high-energy physics
高能质子同步稳相加速器 Bevatron
高频 HF (high frequency)
高频堵漏 high-frequency plugging
高频放电 high-frequency discharge
高频扬声器 tweeter
高熔混合物 dystectic mixture
高渗(压)溶液 hypertonic solution
高斯 gauss
高斯单位 Gaussian units
高斯定理 Gauss(') theorem
高斯定律 Gauss' law
高斯分布 Gaussian distribution
高斯光束 Gaussian beam
高斯光学 Gaussian optics
高斯计 gaussmeter
高斯模型 Gauss model
高斯曲率 Gaussian curvature
高速钢 high-speed steel
高速摄影机 high-speed camera
高速照相机 high-speed camera
高温测量学 pyrometry
高温等离子体 high-temperature plasma
高温计 pyrometer
高压倍加器 Cockcroft-Walton accelerator
高压釜 autoclave
高真空 high vacuum
高真空隔热 high vacuum insulation
高阻尼波 heavy-damped wave
戈 Gy (gray)
戈德斯通定理 Goldstone('s) theorem
戈德斯通粒子 Goldstone particle
戈莱盒 Golay cell
戈瑞 Gy (gray)
哥白尼体系 Copernican system
格波 lattice wave
格波动量 lattice wave momentum
格点 lattice (point)
格点规范理论 lattice gauge theory

格拉姆定律 Graham('s) law
格拉肖-温伯格-萨拉姆模型 GWS model (Glashow-Weinberg-Salam model)
格雷果里望远镜 Gregorian telescope
格林函数 Green function
格栅 grid
格矢 lattice vector
格晕光阑 vignetting stop
格座 lattice site
隔层 interlayer
隔行扫描 alternate-line scanning
隔离 isolation
隔膜 membrane
隔声 sound absorption
隔音度 noise insulation factor
隔音系数 noise insulation factor
镉电池 cadmium cell
个人电脑 PC (personal computer)
个人计算机 PC (personal computer)
各向同性 isotropism; isotropy
各向同性介质 isotropic medium
各向异性 aeolotropism; aeolotropy; anisotropism
各向异性的 anisotropic
各向异性介质 anisotropic medium
各向异性晶体 aeolotropic crystal; anisotropic crystal
各向异性媒质 anisotropic medium
铬钢 chromium steel
根 radical
根式 radical
跟踪 tracking
跟踪雷达 tracking radar
跟踪器 tracker
跟踪制导激光雷达 track-and-guidance laser radar
跟踪装置 tracker
工效学 ergonomics
工作频率 frequency of operation
公差 tolerance
公吨 metric ton
公斤 kg (kilogramme)
公里 km (kilometre)
公顷 hectare
公式 formula
功 work
功函数 work function
功率 power
功率半导体器件 power semiconductor device
功率表 power meter
功率管 power tube
功率耗散 power dissipation
功率计 dynamometer; power meter
功率密度谱 power density spectrum
功率谱 power spectrum
功率三极管 pliotron
功率输出 power output
功率输入 power input
功率速调管 power klystron
功率损耗 power loss
功率消耗 power dissipation
功率因数 power factor
功率整流器 power rectifier
功效函数 power function
功致硬化 work-hardening
汞 mercury
汞灯 mercury lamp
汞合金 amalgam
汞极电池 mercury cell
汞齐 amalgam
汞汽灯 mercury-vapour lamp
汞柱 mercury column
共点力 concurrent force
共轭波 conjugate wave
共轭场 conjugate field
共轭点 conjugate points
共轭光线 conjugate rays
共轭面 conjugate planes
共轭缺陷 conjugate defect
共轭算子 conjugation operator
共轭像 conjugate image
共沸混合物 azeotrope;

azeotropic mixture; constant-boiling mixture
共价键 covalent bond
共价键合 covalent bonding
共价晶体 covalent crystal
共价图 covalent graph
共焦(共振)腔 confocal resonator
共晶混合物 eutectic mixture
共晶转变 eutectic transformation
共面力 coplanar force
共鸣 consonance; resonance
共鸣器 resonator
共生 intergrowth
共吸附 co-adsorption
共线变换 collinear transformation; collineation
共心(共振)腔 concentric resonator
共心光束 homocentric beam
共形变换 conformal transformation
共振 resonance
共振波模 resonant mode
共振电路 resonant circuit
共振峰 resonance peak
共振积分 resonance integral
共振级 resonance level
共振宽度 resonance width
共振模(式) resonant mode
共振能量 resonance energy
共振频率 resonance frequency
共振谱 resonance spectrum
共振器 resonator
共振腔 resonant cavity
共振态 resonance (state)
共振吸收 resonance absorption
共振箱 resonance box
共振荧光 resonance fluorescence
共轴性 coaxiality
沟道效应 channeling effect
构象分析 conformational analysis
构形分析 conformational analysis
构造 configuration; texture
估计 estimation
孤对电子 lone-pair electron
孤立导体 isolated conductor
孤立系 isolated system
孤(立)子 soliton
箍缩效应 pinch effect
古地磁学 palaeomagnetism
钴钢 cobalt steel
鼓风炉 blast furnace
固定 fixing
固定靶 fixed target
固定点 fixed point
固定轮 fast pulley
固定器 fixer
固定指标 fixed index
固溶体 solid solution
固溶线 solvus
固态等离子体 solid-state plasma
固态电路 solid-state circuit
固态探测器 solid-state detector
固态物理 solid-state physics
固体 solid
固体物理 solid-state physics
固相外延 solid phase epitaxy
固有波长 natural wavelength
固有长度 proper length
固有频率 natural frequency
固有时 proper time
固有运动 proper motion
固有振荡 natural oscillation
固有振动 natural oscillation; natural vibration
固有质量 proper mass
固有阻尼 natural damping
故障 defect
拐点 inflection point
拐折 inflection
关联函数 correlation function
关联谱 correlation spectrum
关联系数 correlation coefficient
观测 observation
观测物镜 viewing lens
观测误差 observation error
观测站 observatory

观察 observation
观察孔 observation port
观剧镜 opera glasses
管 tube
贯穿本领 penetrating power
贯穿簇射 penetrating shower
贯穿粒子 penetrating particle
惯量 inertia
惯量积 product of inertia
惯量椭球 ellipsoid of inertia
惯量张量 inertia tensor
惯量轴 axis of inertia
惯量主轴 principal axis of inertia
惯性 inertia
惯性定律 law of inertia
惯性力 inertial force
惯性系 inertial system
惯性系数 inertial coefficient
惯性约束 inertial confinement
惯性张量 inertia tensor
惯性质量 inertial mass
惯性轴 axis of inertia
惯性坐标系 inertial frame
光 light
光泵 light pump
光泵激 optical pumping
光波 light wave
光测高温计 optical pyrometer
光测弹性 photoelasticity
光测弹性学 photoelasticity
光场 light field
光程 optical distance；optical length；optical path
光程差 optical path difference
光抽运 optical pumping
光出射度 luminous exitance
光疵 optical defect
光磁效应 photomagnetic effect
光存储器 optical memory
光带 band
光导 light guide
光导器件 photocon
光导摄像管 vidicon
光导纤维 optical fibre
光导元件 photocon
光的偏振 polarization of light
光电 photoelectricity
光电倍增管 photomultiplier (tube)
光电倍增器 photomultiplier
光电变换器件 photoactor
光电变阻器 photovaristor
光电池 photocell；photoelectric cell；photovoltaic cell
光电传真电报 photoradiogram
光电导 photoconduction
光电导率 photoconductivity
光电导体 photoconductor
光电导效应 photoconductive effect
光电导性 photoconductivity
光电动势 photoelectromotive force
光电二极管 photod；photodiode
光电峰 photopeak
光电管 photocell；photoelectric cell；photoelement；phototube
光电光度计 photoelectric photometer
光电晶体管 phototransistor
光电流 photocurrent；photoelectric current
光电流效应 photogalvanic effect
光电摄像管 iconoscope
光电探测 photodetection
光电探测器 photodetector
光电显像管 iconoscope
光电现象 photoelectricity
光电效应 actinoelectricity；photoeffect；photoelectric effect
光电效应方程 photoelectric equation
光电学 photoelectricity
光电阴极 photocathode
光电阈 photoelectric threshold
光电元件 photocell；photoelectric cell
光电子 photoelectron
光电子发射 photoemission
光电子(能)谱学 photoelectron

spectroscopy
光电子枪 photogun
光电子学 optoelectronics; optronics; photoelectronics
光度测定 photometry
光度计 photometer
光度学 photometry
光度学亮度 photometric brightness
光度佯谬 Olbers' paradox
光二极管 optical diode
光发射 light emission
光阀 light valve
光范性效应 photoplastic effect
光放大 light amplification; optical amplification
光分辨能力 optical resolution
光分解 photolysis
光辐射 optical radiation
光赋色效应 photochromic effect
光赋色性 photochromism
光感受体 photoreceptor
光杠杆 optical lever
光功当量 mechanical equivalent of light
光核反应 photonuclear reaction
光化电离 photoionization
光化度 actinicity
光化反应 photochemical reaction
光化辐射 actinic radiation
光化性 actinicity
光化学 actinology; photochemistry
光混频 photomixing
光活的 photoactive
光激发光 photoluminescence
光激中子 photoneutron
光焦度 focal power
光解 photodissociation
光解作用 photodissociation; photolysis
光阱 light trap
光具盘 optical disk
光具组 optical system
光具座 optical bench
光刻 photoetching; photolithography
光刻法 photolithography
光刻蚀 photoetching
光控场致发光 photoelectroluminescence
光控电致发光 photoelectroluminescence
光控继电器 photoswitch
光控开关 photoswitch
光阑 diaphragm; stop
光缆 optical cable
光亮度 intensity
光量开关 Q-switch
光量子 light quantum
光零差 optical homodyne
光路 path of ray
光路可逆性 reversibility of optical path
光脉冲 light pulse; optical pulse
光脉冲压缩 light pulse compression
光密度计 densitometer
光密介质 optically denser medium
光敏传感器 light sensor
光敏的 photoactive
光敏电阻 LDR (light-dependent resistor); photoconductive resistance; photoresistance; photoresistor; photovaristor
光敏电阻器 LDR (light-dependent resistor); photoresistor; photovaristor
光敏二极管 photodiode
光敏感器 optical sensor
光敏晶体管 phototransistor
光敏三极管 phototransistor
光敏性 photosensitivity
光敏阴极 light-sensitive cathode
光敏元件 light sensor
光敏作用 photosensitization
光能 light energy
光能测定仪 actinometer
光年 light year
光拍 light beat; photo-beat

光盘 CD（compact disk）；optical disk
光盘驱动器 CD–ROM（CD read-only memory）
光偏置 optical biasing
光频差拍 photo-beat
光频声子 optical phonon
光频外差接收器 optical heterodyne receiver
光频外差雷达 optical heterodyne radar
光谱 spectrum
光谱纯度 spectral purity
光谱带 （spectral）band
光谱分析 spectral analysis
光谱级 order of spectrum；spectral order
光谱强度 spectral intensity
光谱术 spectroscopy
光谱特性 spectral characteristic
光谱图 spectrogram
光谱线 spectral line
光谱线系 spectral series
光谱型 spectral type
光谱选择性 spectrum selectivity
光谱学 spectroscopy
光谱仪 spectrometer
光球(层) photosphere
光圈 aperture；diaphragm
光圈号数 f-number
光圈数 focal ratio
光散射 light scattering
光色效应 photochromic effect
光色性 photochromism
光栅 （optical）grating；raster
光栅摄谱仪 grating spectrograph
光渗 irradiation
光生伏打电池 photoelement；photovoltaic cell
光生伏打效应 photovoltaic effect
光声光谱学 photoacoustic spectroscopy
光声效应 optoacoustic effect；photoacoustic effect
光束 light beam；light pencil
光束斑 beam spot
光束半径 beam radius
光束波导 beam waveguide
光束发散度 luminous exitance
光束孔径角 beam angle
光束扫描 optical beam flying
光束腰 beam waist
光束腰半径 waist radius
光速 light velocity；speed of light；velocity of light
光速不变性 constancy of light velocity
光速不变原理 principle of constancy of light velocity
光塑效应 photoplastic effect
光损伤 optical damage
光弹性 photoelasticity
光弹效应 photoelastic effect
光探测 photodetection
光调幅器 optical amplitude modulator
光通量 luminous flux
光通信 optical communication
光瞳 pupil
光瞳函数 pupil function
光透射率 optical transmittance
光外差 optical heterodyne
光微子 photino
光纤 optical fibre
光纤(观察)镜 fibrescope
光纤通信 optical fibre communication
光线 （light）ray
光线追迹 ray tracing
光陷俘 optical trapping
光效应 photoeffect
光心 optical centre
光星 photostar
光行差 aberration
光穴 photohole
光学 optics；photology
光学安培计 optical ammeter
光学变换 optical transform
光学玻璃 optical glass
光学测高仪 optical altimeter

光学测距仪 optical range finder
光学长度 optical length
光学常数 optical constant
光学处理 optical processing
光学传递函数 optical transfer function
光学传感器 optical sensor
光学存储 optical memory
光学电流计 optical ammeter
光学分辨率 optical resolution
光学傅里叶变换 optical Fourier transform
光学高温计 optical pyrometer
光学共振腔 optical resonator
光学恒量 optical constant
光学活性 optical activity
光学畸变 optical distortion
光学集成 optical integration
光学计算机 optical computer
光学介质 optical medium
光学晶体 optical crystal
光学快门 optical shutter
光学零差 optical homodyne
光学媒质 optical medium
光学模 optical mode
光学模拟计算机 optical analogue computer
光学平面 optical flat
光学平行性 optical parallelism
光学器件 optical device
光学全息术 optical holography
光学缺陷 optical defect
光学栅格 optical grating
光学声子 optical phonon
光学双星 optical double star
光学探测器 optical detector
光学天线 optical aerial
光学调准 optical alignment
光学图像 optical image
光学外差 optical heterodyne
光学系统 optical system
光学显微镜 optical microscope
光学陷俘 optical trapping
光学信息处理 optical information processing
光学研磨 optical grinding
光学仪器 optical instrument
光学元件 optical element
光学指数 optical index
光学中心 optical centre
光学周期 optical cycle
光学转移函数 optical transfer function
光学装置 optical device
光学准直 optical alignment
光学(自动)跟踪头 optical homer
光压 light pressure
光以太 luminiferous ether
光阴极 photocathode
光泳 photophoresis
光源 light source
光晕 halo
光增强器 light amplifier
光增益 light amplification
光闸 optical shutter
光制导 optical guidance
光质子 photoproton
光致变色 photochromism
光致变色玻璃 photochromic glass
光致变色性 photochromism
光致电离 photoionization
光致电流效应 photogalvanic effect
光致二向色性 photodichroism
光致发光 photoluminescence
光致分离 photodetachment
光致核反应 photonuclear reaction
光致激发 photoexcitation
光致抗蚀剂 photoresist
光致离解 photodissociation
光致裂变 photofission
光致漂移 photophoresis
光致频移 light shift
光致迁动 photophoresis
光致蜕变 photodisintegration
光致脱离 photodetachment
光致吸附 photoadsorption
光中子 photoneutron
光周期 photoperiod

光轴 optic(al) axis
光轴角 optical angle
光柱 light beam
光锥 light cone；light pencil
光子 photon
光子集束 photon bunching
光子计数 photon counting
光子晶体 photonic crystal
光子能量 photon energy
光子群聚 photon bunching
光子探测器 photon detector
光子通量 photon flux
光子效应 photon effect
光子学 photonics
光阻材料 photoresist
广播 broadcast
广角镜头 orthostigmat；wide-angle lens
广角透镜 pantoscope；wide-angle lens
广角照相机 pantoscope
广延变数 extensive variable
广延量 extensive quantity
广义动量 generalized momentum
广义拉盖尔多项式 generalized Laguerre polynomial
广义朗之万方程 generalized Langevin equation
广义力 generalized force
广义欧姆定律 generalized Ohm law
广义速度 generalized velocity
广义相对论 general theory of relativity
广义相对性原理 principle of general relativity
归一化 normalization
归一(化)因子 normalizing factor
规范变换 gauge transformation
规范玻色子 gauge boson
规范不变性 gauge invariance
规范场 gauge field
规范场论 gauge field theory
规范粒子 gauge particle
规范势 gauge potential
规则反射 regular reflection
硅 silicon
硅二极管 silicon diode
硅光电池 silicon photocell
硅光电二极管 silicon photodiode
硅光电探测器 silicon photodetector
硅铝(明)合金 alpax
硅芯片 silicon chip
轨道 orbit；orbital；trajectory
轨道电子 orbital electron
轨道电子俘获 orbital electron capture
轨道根数 elements of an orbit
轨道量子数 orbital quantum number
轨道速度 orbital speed；orbital velocity
轨道要素 elements of an orbit
轨函数 orbital
轨迹 locus；trajectory
轨迹角 track angle
滚动 rolling motion
滚动摩擦 rolling friction
国际纯粹物理与应用物理联合会 IUPAP (International Union of Pure and Applied Physics)
国际单位制 SI units (Système International d'Unités)
国际符号 international symbol
国际日界线 international date line
国际日期变更线 international date line
国际实用温标 International Practical Temperature Scale
国际烛光 international candle
过饱和 supersaturation
过饱和溶液 supersaturated solution
过饱和蒸汽压 supersaturation vapour pressure
过电压 overvoltage
过电压保护 overvoltage protection；surge protection

过渡层 transition layer
过渡流 Knudsen flow
过渡振动 transient vibration
过冷现象 supercooling phenomenon
过滤 filter；filtration
过滤泵 filter pump
过滤器 filter
过热 overheating
过热电子 hot electron
过热发光 hot luminescence
过剩电子 excess electron
过剩熵 excess entropy
过载 overload
过重 overweight
过阻尼 overdamping

Hh

哈勃常数　Hubble constant
哈勃定律　Hubble law
哈勃天文望远镜　Hubble space telescope
哈佛分类法　Harvard classification
哈雷彗星　Halley's comet
哈密顿函数　Hamiltonian (function)
哈密顿量　Hamiltonian
哈密顿算符　Hamiltonian (operator)
哈密顿特征函数　Hamiltonian characteristic function
哈密顿原理　Hamilton principle
海拔　altitude
海拔高度　height above sea level
海里　nautical mile
海森伯测不准原理　Heisenberg uncertainty principle
海市蜃楼　mirage
海啸　seaquake
海震　seaquake
亥姆霍兹方程　Helmholtz equation
亥姆霍兹-拉格朗日定理　Helmholtz-Lagrange theorem
亥姆霍兹线圈　Helmholtz coils
亥姆霍兹自由能　Helmholtz free energy
亥维赛层　Heaviside layer
亥维赛-洛伦兹单位　Heaviside-Lorentz units
氦镉激光器　helium-cadmium laser
氦冷冻机　helium refrigerator
氦氖激光器　He-Ne laser (helium-neon laser)
氦制冷机　helium refrigerator
含氘化合物　deuterated compound
含水的　aqueous
含水晶体　hydrated crystal
含陶合金　cermet
函数　function
焓　enthalpy
汉弗莱级数　Humphrey series
焊滴结　solder drop junction
焊剂　flux
行式打印机　line printer
航空弹道学　aeroballistics
航空地磁仪　airborne magnetometer
航空电子学　avionics
航空器　aircraft
航空声学　aeroacoustics
航空学　aeronautics
航空照相机　aerocamera
航摄相片　airphoto
航天学　astronautics
毫安表　milliammeter
毫安计　milliammeter
毫巴　mB (millibar)
毫居里　mC (millicurie)
毫克　mg (milligram)
毫米　mm (millimetre)
毫米汞柱　mmHg
毫升　ml (millilitre)
毫微米　nm (nanometre)
好量子数　good quantum number
郝斯勒合金　Heusler alloys
耗尽层　depletion layer
耗尽效应　depletion effect
耗散　dissipation

耗散结构 dissipative structure
耗散力 dissipative force
耗散流 dissipative flow
耗散系统 dissipative system
合 conjunction
合成孔径 synthetic aperture
合成色 secondary colour
合金 alloy
合金钢 alloy steel
合金结 alloyed junction
合金结晶体管 alloyed junction transistor
合理单位 rationalized unit
合力 composition of forces; resultant (force)
合力偶 resultant couple
合量 resultant
合速度 resultant velocity
合轴组 centred system
合作发光 cooperative luminescence
合作发射 cooperative emission
合作现象 cooperative phenomenon
河外星系 external galaxy
核 nucleus
核半径 nuclear radius
核磁共振 NMR (nuclear magnetic resonance)
核磁性 nuclear magnetism
核磁子 nuclear magneton
核弹头 nuclear warhead
核电荷 nuclear charge
核电荷数 nuclear charge number
核电站 nuclear power station
核动力 nuclear power
核反应 nuclear reaction
核反应堆 nuclear reactor
核废料 nuclear waste
核分裂 nuclear fission
核辐射 nuclear radiation
核共振 nuclear resonance
核合成 nucleosynthesis
核结构 nuclear structure
核矩 nuclear moment
核聚变 atomic fusion; nuclear fusion
核壳层模型 nuclear shell model
核扩散 nuclear proliferation
核力 nuclear force
核粒子 nuclear particle
核裂变 nuclear fission
核裂产物 fission product
核模型 nuclear model
核能 atomic energy; nuclear energy
核谱学 nuclear spectroscopy
核燃料 nuclear fuel
核热 nuclear heat
核衰变 nuclear decay
核顺磁共振 NPR (nuclear paramagnetic resonance)
核素 nuclide
核素图 chart of nuclides
核同质异能性 nuclear isomerism
核武器 nuclear weapon
核物理学 nuclear physics
核心 core
核心极化 core polarization
核影散射 shadow scattering
核战争 nuclear war
核质点 nuclear particle
核质量 nuclear mass
核质子 nuclear proton
核装置 nuclear device
核子 nucleon
核子核心 nucleor
核子偶素 nucleonium
核子数 mass number; nucleon number
核子学 nucleonics
核自旋 nuclear spin
荷电时间常量 charging time constant
荷质比 charge-mass ratio; specific charge
赫罗图 H-R diagram (Hertzsprung-Russell diagram)
赫斯定律 Hess's law
赫兹 Hz (hertz)
赫兹振子 Hertzian oscillator

黑白群 black-and-white group
黑洞 black hole
黑体 black body
黑体辐射 black-body radiation
痕量金属 tracer metal
亨利 H (henry)
亨利定律 Henry's law
哼鸣 hum
哼扰 hum
恒等(式) identity
恒定场 stationary field
恒定电流 steady current
恒沸物 azeotrope; azeotropic mixture; constant-boiling mixture
恒力 constant force
恒流源 constant current source
恒温器 thermostat
恒温箱 thermostat
恒星 (fixed) star
恒星风 stellar wind
恒星干涉仪 stellar interferometer
恒星日 sidereal day
恒星视差 stellar parallax
恒星视亮度 apparent stellar brightness
恒星演化 stellar evolution
恒星中天摄影术 photochronograph
恒星周期 sidereal period
恒星状球体 quasar
恒压源 constant voltage source
横波 transverse wave
横电波 transverse electric wave
横断面 cross-section; transverse section
横截面 cross-section; transverse section
横力 transversal force
横脉冲 transverse impulse
横能量 transverse energy
横向弛豫 transverse relaxation
横向倒置 lateral inversion; perversion
横向放大率 lateral magnification
横向分量 transverse component
横向负载 transverse load
横向加速度 transverse acceleration
横向扩散 transverse diffusion
横向平移 transverse translation
横向强度 transverse strength
横向速度 lateral velocity; transverse velocity
横向相干性 lateral coherence; transverse coherence
横向像差 lateral aberration
横向应力 transverse stress
横向照度 transverse illumination
横向照明 transverse illumination
横振动 transverse vibration
横质量 transverse mass
横轴 transverse axis
横坐标 abscissa
衡消法 null method
轰击粒子 bombarding particle
红矮星 red dwarf
红宝石激光器 ruby laser
红巨星 red giant
红外的 ultrared
红外灯 infrared lamp
红外电子学 infranics
红外发散 infrared divergence
红外分光光度计 infrared spectrophotometer
红外辐射 IR (infrared radiation); nocturnal radiation
红外光谱学 infrared spectroscopy
红外激光器 infrared laser
红外受激发光 infrared stimulated luminescence
红外天文学 infrared astronomy
红外线 infrared ray; ultrared
红外线温度计 infrared thermometer
红移 redshift
宏观的 macroscopic
宏观极化 macroscopic polarization

宏观结构 macrostructure
宏观量 macroscopic quantity
宏观量子现象 macroscopic quantum phenomenon
宏观态 macroscopic state
宏观因果性 macrocausality
宏晶 macrocrystal
虹膜 iris
虹霓 rainbow
虹吸(管) siphon
洪德定则 Hund's rules
后备存储器 backing store
后处理 aftertreatment
弧度 circular measure; rad (radian)
弧度法 circular measure
弧光灯 arc lamp
弧光放电 arc discharge
弧矢焦线 sagittal focal line
胡克定律 Hooke's law
蝴蝶效应 butterfly effect
互补色 complementary colour
互补型金属氧化物半导体 CMOS (complementary metal-oxide semiconductor)
互补性 complementarity
互补原理 complementary principle
互导 mutual conductance
互感 mutual induction
互感器 mutual inductor
互感系数 mutual inductance
互感应 mutual induction
互联网 Internet
互能 mutual energy
互相干性 mutual coherence
护真光阑 vignetting stop
华氏度 degree Fahrenheit
华氏温标 Fahrenheit (temperature) scale; Fahrenheit (thermometric) scale
华氏温度计 Fahrenheit
滑车组 block and tackle
滑动光学计算尺 variopter
滑动摩擦 sliding friction
滑动摩擦系数 coefficient of sliding friction
滑轮 pulley; wheel
滑轮组 block and tackle; pulley blocks
滑线电桥 metre bridge
滑移带 slip band
滑移反射 glide reflection
滑移面 glide plane
化合光 chemiluminescence
化合价 valence; valency
化合物 combination; compound
化学成键 chemical bonding
化学电池 chemical cell
化学电离 chemi-ionization
化学电源 electrochemical cell
化学发光 chemiluminescence
化学分析电子能谱法 ESCA (electron spectroscopy for chemical analysis)
化学光度计 actinometer
化学键 chemical bond
化学能 chemical energy
化学汽相沉积 CVD (chemical vapour deposition)
化学亲和势 chemical affinity
化学亲和性 chemical affinity
化学势 chemical potential
化学物理学 chemical physics
化学吸附 chemisorption
化学荧光 chemiluminescence
划分 partition
话筒 microphone
环 annulus; loop; torus
环量 circulation
环流 circulation
环路 ring circuit; ring main
环面透镜 toric lens
环球网 Web; WWW (World Wide Web)
环食 annular eclipse
环形电路 ring circuit; ring main
环形模 ring mode
环形物 annulus
缓变结 graded junction
缓变异质结 graded

heterojunction
缓冲电路 buffer circuit
缓冲寄存器 buffer register
缓动装置 decelerator
缓发质子 delayed proton
缓发中子 delayed neutron
缓和剂 moderator
幻核 magic nucleus
幻数 magic number
换能器 transducer；transverter
换算因子 conversion factor
换算装置 scaler
换向器 commutator
换相器 phase changer
黄道 ecliptic；zodiac
黄道带 zodiac
黄道(带)光 zodiacal light
黄铜 brass
簧片开关 reed switch
灰阶 grey level
灰(色)群 grey group
恢复系数 coefficient of restitution；restitution coefficient
辉光 glow
辉光放电 glow discharge
辉光放射 airglow emission
辉光曲线 glow curve
回摆晶体 oscillating crystal
回波 echo
回复力 restoring force
回归分析 analysis of regression；regression analysis
回归年 solar year；tropical year
回火 tempering
回路 circuit；loop
回声 echo
回声测定 echolocation
回声定位法 echolocation
回声室 echo chamber
回声探测器 echo sounder
回弹系数 coefficient of restitution；restitution coefficient
回跳硬度 scleroscope hardness
回跳硬度计 sclerometer；scleroscope
回响室 echo chamber
回旋半径 cyclotron radius；gyroradius；radius of gyration
回旋不稳定性 cyclotron instability
回旋辐射 cyclotron radiation
回旋共振 cyclotron resonance
回旋加速器 cyclotron
回旋频率 cyclotron frequency
回旋质谱仪 omegatron
回旋装置 gyroscope
回音 echo
回音测深器 echo sounder
回音测深仪 fathometer；sonic depth finder
回转罗盘 gyrocompass
回转仪 gyroscope
回转轴 rotor
会合周期 synodic period
会聚波 convergent wave
会聚镜 convergent mirror
会聚透镜 convergent lens
会切场 cusp field
彗发 coma
彗星 comet
彗形像差 coma；comatic aberration
惠更斯原理 Huygens(') principle
惠更斯作图法 Huygens(') construction
惠斯通电桥 Wheatstone bridge
混沌 chaos
混沌测度 chaotic measure
混沌场 chaotic field
混沌激光 chaotic laser light
混沌运动 chaotic motion
混合 admixture
混合法 method of mixtures
混合器 mixer
混合态 mixed state
混合物 admixture；compound
混合性 mixing
混晶 mixed crystal
混频 (frequency) mixing

混频器 mixer
混响 reverberation
混响声 reverberant sound
混响时间 reverberation time
混杂电荷 heterocharge
混杂子 hybrid
活动区 active region
活动物体照相机 photochronograph
活动星系 active galaxy
活度 activity
活化 activation
活化分析 activation analysis
活化剂 activating agent
活化能 activation energy
活化势 activation potential
活化态 activated state
活化原子 activated atom
活塞 piston
活塞杆 piston rod
活塞运动 piston motion
活性 activity
活性核 active nucleus
活性区 active section
活性衰减 activity decay
活性炭 activated carbon
火花淬灭 spark quenching
火花电蚀 spark erosion
火花放电 spark discharge
火花光谱 spark spectrum
火花计数器 spark counter
火花塞 spark plug
火花室 spark chamber
火花隙 spark gap
火箭 rocket
火焰 flame
火焰光度计 flame photometer
或电路 OR circuit
或非门 NOR gate
霍耳效应 Hall effect

Jj

几率　probability
几率函数　probability function
几率积分　probability integral
几率密度　probability density
击波管　shock tube
击穿　breakdown
击穿场强　breakdown field strength
击穿电压　breakdown voltage
击打式打印机　impact printer
机构系统　train of mechanism
机理　mechanism
机器人学　robotics
机械波　mechanical wave
机械功　mechanical work
机械利益　MA（mechanical advantage）
机械能　mechanical energy
机械能守恒　conservation of mechanical energy
机械能守恒定律　law of conservation of mechanical energy
机械系　mechanical system
机械运动　mechanical motion
机械振动　mechanical vibration
机械致热效应　mechanocaloric effect
机械装置　mechanism
机翼　aerofoil；airfoil
机载磁强计　airborne magnetometer
机载激光跟踪器　ALT（airborne laser tracker）
奇奇核　odd-odd nucleus
奇偶核　odd-even nucleus
奇偶检验　parity check
奇偶校验位　parity bit
奇偶性　parity
奇宇称　odd parity
积分常数　constant of integration
积分法　integration
积分路线　path of integration
积分学　integral calculus
基本常量　fundamental constant
基本常数　fundamental constant
基本单位　base unit；fundamental unit
基本光束　elementary beam
基本粒子　elementary particle；ultimate particle
基本原理　fundamental
基本振动　fundamental vibration
基波　fundamental wave；principal wave
基点　cardinal point
基尔霍夫定律　Kirchhoff's laws
基尔霍夫辐射定律　Kirchhoff's law of radiation
基极　base
基面　cardinal plane
基模　fundamental mode
基频　fundamental frequency
基区掺杂　base doping
基数　base
基态　ground state
基线系　fundamental series
基谐波　first harmonic
基谐模　fundamental mode
畸变　deformation；distortion
激发　activation；excitation
激发函数　excitation function
激发能　excitation energy

激发曲线 excitation curve
激发态 excited state
激发子 exciton
激光 laser
激光泵 laser pump
激光测距仪 laser range finder
激光抽运 laser pumping
激光打孔 laser boring
激光打印机 laser printer
激光二极管 laser diode
激光防撞雷达 obstacle avoidance laser radar
激光放大器 laser amplifier
激光功率计 laser power meter
激光共振腔 laser cavity
激光光谱 laser spectrum
激光焊接 laser bonding
激光基质 laser host
激光激发 laser excitation
激光尖峰 spike
激光校直 laser aligning
激光雷达 lidar (light detection and ranging)
激光模(式) mode of laser
激光器 laser; optical maser
激光器基质 laser host
激光腔 laser cavity
激光切割 laser cutting
激光束 laser beam
激光退火 laser annealing
激光陀螺 laser gyroscope
激光物理学 laser physics
激光阵列 laser array
激光振荡器 laser oscillator
激光制导 laser guidance
激光制冷 laser cooling
激活 activation
激活电子 active electron
激活分析 activation analysis
激活光纤 active optical fibre
激活剂 activating agent
激活介质 active medium
激活能 activation energy
激活能级 active level
激活腔 active cavity
激活区 active region
激活态 activated state
激活原子 activated atom
激活载流子 activated carrier
激活载体 activated carrier
激基分子激光器 excimer laser; exciplex laser
激声 phaser
激子 barion; exciton
激子光谱 exciton spectrum
激子模型 exciton model
吉 billion
吉伯 Gb (gilbert)
吉布斯函数 Gibbs function
吉布斯佯谬 Gibbs paradox
吉布斯自由能 Gibbs free energy
吉咖 billion
汲取 pump
级 degree; order
级联变压器 cascade transformer
级联发射 cascade emission
级联放射 cascade radiation
级联辐射 cascade radiation
级联过程 cascade process
级联激光器 cascade laser
级联冷却 cascade cooling
级联衰变 cascade decay
级联跃迁 cascade transition
级数 series
极 pole
极板 plate
极低频 ELF (extremely low frequency)
极点 pole
极端光程 extreme path
极端紫外 extreme ultraviolet
极高频 EHF (extremely high frequency)
极光 aurora
极化 polarization
极化电荷 polarization charge
极化电流 polarization current
极化度 polarizability; polarization
极化方向 polarization direction
极化率 polarizability
极化曲线 polarization curve

极化效应 polarization effect
极化性 polarizability
极化轴 polaxis
极化子 polaron
极距 polar distance；pole distance
极谱记录仪 polarograph
极强 pole strength
极矢（量） polar vector
极隧射线 canal ray
极隙 cusp
极限 limit
极限环 limit cycle
极限能量 ultimate energy
极限强度 ultimate strength
极限群 limiting group
极限速度 limiting velocity
极性 polarity
极性分子 polar molecule
极性键 polar bond
极性晶体 polar crystal
极靴 pole piece
极远光程 extreme path
极远紫外 extreme ultraviolet
极柱 terminal
极坐标 polar coordinates
即时速度 instantaneous velocity
即时中心 instantaneous centre
急冷 rapid cooling
急流 jet stream
集成 integration
集成电路 IC（integrated circuit）；silicon chip
集成光路 integrated optical circuit
集成光学 integrated optics
集电极 collector
集光器 collector
集合 set
集合论 set theory
集体激发 collective excitation
集体拍 collective beating
集体频率 collective frequency
集体振荡 collective oscillation
集体振动 collective vibration
集体转动 collective rotation
集团 cluster
几何光学 geometrical optics
几何化单位 geometrized units
几何级数 geometric series
几何平均数 geometric average
几何像差 geometrical aberration
计 meter
计量学 metrology
计时电路 timing circuit
计时器 timer
计数电路 scaling circuit
计数率 counting rate
计数器 counter
计算机 computer
计算机层析成像 CT（computerized tomography）
计算机辅助设计 CAD（computer-aided design）
计算机轴向层析成像 CAT（computerized axial tomography）
计算物理学 computational physics
计温学 thermometry
剂量 dose
剂量当量 dose equivalent
剂量计 dosimeter
迹线 trajectory
继电器 relay
寄生电容 parasite capacitance
寄生反馈 parasitic feedback
寄生反射器 parasitic reflector
寄生俘获 parasitic capture
寄生共振 parasitic resonance
寄生散射 parasitic scattering
寄生损失 parasitic loss
寄生现象 parasitics
寄生像 parasitic image
寄生效应 parasitic effect；parasitics
寄生元件 parasitic element
寄生振荡 hunting；parasitic oscillation
寄生振荡遏抑器 parasitic suppressor
加法放大器 adder amplifier

加法混色 additive colour mixing
加法器 adder
加工硬化 work-hardening
加仑 gal (gallon)
加拿大胶 Canada balsam
加色法 additive process
加速参照系 accelerating reference frame
加速度 acceleration
加速度测量仪 accelerometer
加速度计 accelerometer
加速器 accelerator
加速运动 accelerated motion
加性定律 additivity law
加压舱 pressurized cabin
加压釜 autoclave
假设 hypothesis
假说 hypothesis
价 valence；valency
价带 valence band
价电荷 valency charge
价电子 valence electron
价电子层 valence shell
价电子带 valence band
价核子 valence nucleon
价壳层 valence shell
价力 valence force
价态 valence state
尖点 cusp
尖端 point
尖端放电 point discharge
尖峰 spike
尖峰脉冲 spike pulse
尖拐(型突变) cusp
检波 demodulation；detection
检波器 (wave) detector
检波效率 detection efficiency
检测 detection
检流计 galvanometer
检偏器 analyzer
减反射膜 antireflecting film
减幅 damping
减色法 subtractive process
减湿器 dehumidifier
减速 deceleration；retardation
减速板 airbrake；retardation plate
减速度 deceleration
减速火箭 retrorocket
减速剂 moderator
减速器 decelerator；speed reducer
剪力 shearing force
剪切干涉仪 shearing interferometer
剪切模量 shear modulus
剪应变 shearing strain
简并半导体 degeneracy semiconductor；degenerate semiconductor
简并度 degeneracy
简并量子气体 degenerate quantum gas
简并能级 degenerate energy level
简并气体 degenerate gas
简并态 degenerate states
简并温度 degeneracy temperature
简并系统 degenerate system
简并性 degeneracy
简并性判据 degeneracy criterion
简并压力 degeneracy pressure
简谐波 simple harmonic wave
简谐运动 SHM (simple harmonic motion)
简正模(式) normal mode
简正频率 normal frequency
简正振动 normal (mode of) vibration
简正坐标 normal coordinate
碱 alkali
碱度 alkalinity
碱金属 alkaline metal
碱金属光谱 alkali spectrum
碱金属锑化物 alkali-antimonides
碱性 alkalinity
间壁 partition
间断函数 discontinuous function
间隔 space
间接测量 indirect measurement
渐近线 asymptote

渐近自由 asymptotic freedom
渐开线 involute
渐屈线 evolute
渐伸线 involute
渐晕光阑 vignetting stop
溅射 spattering; sputtering
鉴别 discrimination
鉴别器 discriminator
鉴频器 discriminator
键长 bond distance
键合 bonding
键距 bond distance
键能 bond energy
江崎二极管 Esaki diode; tunnel diode
降频转换 frequency down-conversion
降压变压器 step-down transformer
交变磁场 alternating magnetic field
交变磁通量 alternating flux
交变电场 alternating electric field
交变电动势 alternating electromotive force
交变电压 alternating voltage
交变通量 alternating flux
交叠〖级的〗 overlapping of orders
交互生长 intergrowth
交互式 CD CD-I (CD interactive)
交互作用 interaction
交换对称性 exchange symmetry
交换力 exchange force
交换律 commutative law
交换能 exchange energy
交换群 Abelian collection; Abelian group
交换作用 exchange interaction
交流(电) AC; a. c. (alternating current)
交流电场 alternating electric field
交流电路 alternating circuit
交流电桥 a. c. bridge
交流电压 alternating voltage
交流发电机 alternator
交流放大器 alternating-current amplifier
交流输出 alternating-current output
交流自整角机 teletorque
交直流两用电动机 universal motor
胶合 cementation
胶合双透镜 cemented doublet
胶片剂量计 film badge
胶片式射线计量器 film badge
胶球 glueball
胶体 colloid
胶微子 gluino
胶质 colloid
胶子 gluon
焦比 focal ratio
焦点 focal point; focus
焦点容限 focal tolerance
焦度 (focal) power
焦耳 J (joule)
焦耳定律 Joule's laws
焦耳加热 Joule heating
焦耳-开尔文效应 Joule-Kelvin effect
焦耳热 Joule heat
焦耳实验 Joule experiment
焦耳-汤姆孙系数 Joule-Thomson coefficient
焦耳-汤姆孙效应 Joule-Thomson effect
焦距 focal length
焦距计 focometer
焦阑 telecentric stop
焦阑的 telecentric
焦阑系统 telecentric system
焦(平)面 focal plane
焦散的 caustic
焦散线 caustics
焦深 depth of focus
角动量 angular momentum; moment of momentum
角动量守恒 angular-momentum

conservation; conservation of angular momentum
角动量守恒定律 angular-momentum conservation law; law of conservation of angular momentum
角反射镜 angle mirror
角放大率 angular magnification; angular magnifying power
角分辨率 angular resolution
角加速度 angular acceleration
角镜 angle mirror
角距 angular distance
角孔径 angular aperture
角量子数 angular quantum number; azimuthal quantum number
角频率 angular frequency; pulsatance
角色散 angular dispersion
角速度 angular speed; angular velocity
角调制 angle modulation
角位移 angular displacement
角向磁场 poloidal magnetic field
角运动 angular motion
角锥 pyramid
绞合线 twisted wire
矫顽电场 coercive electric field
矫顽力 coercive force; coercivity
矫顽性 coercivity
矫顽应力 coercive stress
校正 correction
校正板 correction plate
校正不完全透镜 undercorrected lens
校正场 correcting field
校正透镜 correcting lens
校直 straightening
校准电导水 conductivity water
校准器 aligner
阶 order
阶乘(积) factorial
阶式减光板 stepped attenuator
阶梯光栅 echelon grating
阶梯近似 ladder approximation
阶梯脉冲 stepped pulse; step pulse
阶梯绕组 stepped winding
阶梯透镜 stepped lens
接触电势差 contact potential difference
接触力 contact force
接地 earth; ground
接地线 earth wire
接目镜 eye lens
接受器 acceptor
接线柱 post
节点 nodal point; node
节点滑座 nodal slide
节流阀 throttle valve
洁净表面 clean surface
结点 node
结构 mechanism
结构弛豫 structural relaxation
结构函数 structure function
结构相变 structural phase transition
结构因子 structure factor
结合 coupling
结合律 associative law
结合能 binding energy
结晶 crystal
结晶度 crystallinity
结晶固体 crystalline solid
结晶学 crystallography
结栅场效应晶体管 JUGFET (junction gate field effect transistor)
结型场效应管 JFET (junction field effect transistor)
结型检波器 junction detector
结型晶体管 junction transistor
睫状体 ciliary body
截断 cut
截光器 light chopper
截面 cross-section
截止波长 cutoff wavelength
截止频率 cutoff frequency

解理 cleavage
解调 demodulation
解吸(作用) desorption
解析函数 analytic function
解析几何学 analytical geometry
介电常量 dielectric constant; permittivity
介电常数 dielectric constant; permittivity
介电各向异性 dielectric anisotropy
介电极化 dielectric polarization
介电极化率 dielectric susceptibility
介电老化 dielectric aging
介电强度 dielectric strength
介电损耗 dielectric loss
介电损失 dielectric loss
介电陶瓷 dielectric ceramics
介电体 dielectric
介电体击穿 dielectric breakdown
介观结构 mesoscopic structure
介观物理学 mesoscopic physics
介质 medium
介质强度 dielectric strength
介子 meson
界层 interlayer
界面 interface
界面波 capillary wave
界面层 boundary layer
界面态 interface state
界限 threshold
金刚石 diamond
金相学 metallography
金属 metal
金属-半导体接触 metal-semiconductor contact
金属玻璃 metallic glass; metglass
金属点阵 metallic lattice
金属光栅 metallic lattice
金属互化物 intermetallic compound
金属灰 calx
金属键合 metallic bonding
金属-绝缘体-半导体结构 metal-insulator-semiconductor structure
金属-绝缘体转变 metal-insulator transition
金属疲劳 metal fatigue
金属陶瓷 cermet; metallic ceramics
金属物理学 metal physics
金属-氧化物-半导体场效应晶体管 MOSFET (metal-oxide-semiconductor field effect transistor)
金属-氧化物-半导体结构 metal-oxide-semiconductor structure
金属-氧化物-半导体晶体管 MOS transistor (metal-oxide-semiconductor transistor)
金属组织学 metallography
紧束缚近似 tight-binding approximation
进动 precession; precessional motion
进气阀 air admission valve
近场图样 near field pattern
近代物理学 modern physics
近地点 perigee
近点 near point; periapsis
近点角 anomaly
近点距 anomaly
近拱点 periapsis
近红外 near infrared
近晶相 smectic phase
近日点 perihelion
近视 myopia; nearsightedness; short-sightedness
近似技术 approximation technique
近星点 periastron
近月点 pericynthion; perilune
近轴的 paraxial
近轴光线 near(ly) axial ray
近轴区 paraxial region
近紫外 near ultraviolet
浸渐过程 adiabatic process
浸渐近似 adiabatic

approximation
浸渐条件 adiabatic condition
浸没物镜 immersion objective
禁带 forbidden band
禁戒发射 forbidden emission
禁戒跃迁 forbidden transition
经典场论 classical field theory
经典电动力学 classical electrodynamics
经典电子半径 classical electron radius
经典光学 classical optics
经典金属电子论 classical electron theory of metal
经典力学 classical mechanics; Newtonian mechanics
经典统计法 classical statistics
经典物理学 classical physics
经典相干性 classical coherence
经典自旋模型 classical spin model
经典自由度 classical degree of freedom
经度 longitude
经纬测角仪 pantometer
经纬仪 theodolite
经线 longitude
经验公式 empirical formula
经验物态方程 empirical equation of state
晶胞 (unit) cell
晶胞大小 unit cell dimension
晶带指数 zone index
晶带轴 zone axis
晶格 (crystal) lattice
晶格常量 lattice constant
晶格动力学 lattice dynamics
晶格理论 lattice theory
晶格能 lattice energy
晶格热容 lattice heat capacity
晶格散射 lattice scattering
晶格失配 lattice mismatch
晶格振动 lattice vibration
晶格振动吸收 lattice vibrational absorption
晶核 crystal nucleus
晶核生成 nucleation
晶界 grain boundary
晶类 crystal class
晶棱 crystal edge
晶粒 crystalline grain
晶列 crystal column
晶面 crystal face
晶面矢 crystal plane vector
晶面因数 planar factor
晶面指数 indices of crystal plane
晶片 chip; wafer
晶溶发光 lyoluminescence
晶态 crystalline state
晶体 crystal
晶体传声器 crystal microphone
晶体点阵 crystal lattice
晶体管 transistor
晶体管等效电路 transistor equivalent circuit
晶体管电路 transistor circuit
晶体管放大器 transistor amplifier
晶体管-晶体管逻辑 TTL (transistor-transistor logic)
晶体管物理学 transistor physics
晶体惯态 crystal habit
晶体光谱 crystal spectrum
晶体光学 crystal optics
晶体计数器 crystal counter
晶体结构 crystal(line) structure
晶体缺陷 crystal defect
晶体生长 crystal growth
晶体拾音器 crystal pick-up
晶体物理学 crystal physics
晶体学 crystallography
晶体衍射图 crystallogram
晶体振荡器 crystal oscillator; piezoelectric oscillator
晶系 crystal system; syngony
晶向 crystal direction
晶向矢 crystal direction vector
晶向指数 indices of crystal direction
晶形 crystal form
晶性 crystallinity

晶缘 crystal edge
晶闸管 thyristor
晶轴 crystallographic axis
晶轴角 crystallographic axial angle
精密度 precision
精密校正 fine adjustment
精密仪器 precise instrument
精细结构 fine structure
景深 depth of field
劲度 stiffness
劲度模量 stiffness modulus
劲度系数 coefficient of stiffness; stiffness coefficient
径迹 track
径迹长度 track length
径迹角 track angle
径迹室 track chamber
径迹衰退 track fading
径迹退化 track fading
径量子数 radial quantum number
径矢 radius vector
径向动量 radial momentum
径向分布函数 radial distribution function
径向加速度 radial acceleration
径向结构函数 radial structure function
径向量 radius vector
径向扫描 radial sweep
径向速度 radial velocity
净电荷 net charge
净化 cleaning
净力 net force
静长度 proper length
静磁学 magnetostatics
静电 static electricity
静电场 electrostatic field
静电场环路定理 circuital theorem of electrostatic field
静电除尘器 electrostatic precipitator
静电传声器 capacitor microphone; condenser microphone
静电发生器 electrostatic generator
静电复印 xerography
静电感应 electrostatic induction
静电积累 electrostatic accumulation
静电计 electrometer
静电加速器 Van de Graaff accelerator
静电聚焦 electrostatic focusing
静电离子显微镜 FIM (field-ionization microscope; field-ion microscope)
静电力 electrostatic force
静电能 electrostatic energy
静电平衡 electrostatic equilibrium
静电屏蔽 electrostatic screening; electrostatic shielding
静电透镜 electrostatic lens
静电系单位 e.s.u. (electrostatic units)
静电泄漏 electrostatic leakage
静电学 electrostatics; static electricity
静电灾害 electrostatic hazard
静力学 statics
静粒子 stationary particle
静摩擦 static friction
静摩擦系数 coefficient of static friction
静能 rest energy
静态寄存器 static register
静胁强 static stress
静应力 static stress
静噪滤波器 noise filter
静止轨道 stationary orbit
静止质量 proper mass
静质量 rest mass
镜顶 vertex of mirror
镜(面)反射 mirror reflection; regular reflection; specular reflection
镜铁 spiegel(eisen)
镜头 lens

镜像 mirror image
镜像法 method of images
镜像反射 mirror reflection
镜像核 mirror nuclei
镜子 mirror
纠正 rectification
久期方程 characteristic equation; secular equation
酒石酸钾晶体 Rochelle salt
居间键 intermediate bond
居间耦合 intermediate coupling
居间态 intermediate state
居间相 intermediate phase
居间像 intermediate image
居里 curie
居里点 Curie point; Curie temperature
居里定律 Curie's law
居里温度 Curie point; Curie temperature
局部化 localization
局部理论 local theory
矩 moment
矩阱势 rectangular-well potential
矩阵 matrix
矩阵代数 algebra of matrices
矩阵力学 matrix mechanics
巨配分函数 grand partition function
巨势 grand potential
巨星 giant star
巨正则分布 grand canonical distribution
巨正则系综 grand canonical ensemble
距离 distance
距离比 distance ratio
锯齿波发生器 sawtooth generator
锯齿波扫描 sawtooth sweep
锯齿波形 sawtooth waveform
聚变 fusion
聚变反应 fusion reaction
聚变反应堆 fusion reactor
聚光镜 optical collector
聚光器 condenser; optical collector
聚合度 degree of polymerization
聚合体 polymer
聚合物 polymer
聚合物半导体 polymer semiconductor
聚焦 focus
聚焦电极 focusing electrode
聚声器 sound concentrator
聚星 multiple star
绝对参照系 absolute reference frame
绝对单位制 c. g. s. system
绝对电容率 absolute permittivity
绝对伏特 abvolt
绝对黑体 absolute black body; perfect black body
绝对活度 absolute activity
绝对加速度 absolute acceleration
绝对零度 absolute zero
绝对欧姆 abohm
绝对膨胀 absolute expansion
绝对湿度 absolute humidity
绝对时间 absolute time
绝对速度 absolute velocity
绝对位移 absolute displacement
绝对温标 absolute scale of temperature; absolute temperature scale; Kelvin (temperature) scale
绝对温度 absolute temperature
绝对误差 absolute error
绝对音调 absolute pitch
绝对运动 absolute motion
绝对值 absolute value; actual value
绝热 heat insulation; thermal insulation
绝热方程 adiabatic equation
绝热方程〖理想气体的〗 adiabatic equation of ideal gas
绝热过程 adiabatic process
绝热霍尔效应 adiabatic Hall

effect
绝热膨胀 adiabatic expansion
绝热器 （thermal）insulator
绝热式近似 adiabatic approximation
绝热条件 adiabatic condition
绝热图 adiabatic chart；adiabatic diagram
绝热退磁 adiabatic demagnetization
绝热线 adiabat
绝热压缩 adiabatic compression
绝热压缩加热 adiabatic compression heating
绝热指数 adiabatic exponent
绝缘 insulation；isolation
绝缘材料 insulant
绝缘介质 insulating medium
绝缘强度 dielectric strength
绝缘栅场效(应)管 IGFET（insulated-gate field effect transistor）
绝缘体 insulator；isolator
倔强系数 coefficient of stiffness
倔强性 stiffness
倔强因数 stiffness factor
均方位移 mean square displacement
均相外延 homoepitaxy
均匀电介质 uniform dielectric
均匀照明 uniform illumination

Kk

喀戎星　Chiron
卡鲁扎-克莱因理论　Kaluza-Klein theory
卡路里　calorie
卡诺定理　Carnot theorem
卡诺效率　Carnot efficiency
卡诺循环　Carnot cycle
卡塞格林望远镜　Cassegrain telescope
卡西米尔效应　Casimir effect
开　K（kelvin）
开度角　opening angle
开尔文温标　Kelvin（temperature）scale
开尔文效应　Kelvin effect
开方　evolution；extraction
开放系统　open system
开关　switch
开关时间　switching time
开金　carat；karat
开路　open circuit
开普勒定律　Kepler's laws
开氏温标　Kelvin（temperature）scale
开系　open system
开宇宙　open universe
凯特摆　Kater's pendulum
坎（德拉）　cd（candela）
康普顿波长　Compton wavelength
康普顿电子　Compton electron
康普顿散射　Compton scattering
康普顿效应　Compton effect
康铜　constantan
抗磁力　coercive force
抗磁性　diamagnetism
抗断强度　breaking strength
抗拉强度　tensile strength
抗挠强度　bending strength；transverse strength
抗扭劲度　torsional rigidity
抗扭强度　torsional strength
抗扰度　noise immunity
抗扰性　noise immunity
抗弯强度　bending strength
抗张强度　tensile strength
柯西色散公式　Cauchy dispersion formula
科尔劳希平方根定律　Kohlrausch square root law
科克罗夫特-瓦尔顿加速器　Cockcroft-Walton accelerator
科里奥利加速度　Coriolis acceleration
科里奥利力　Coriolis force
科特雷尔静电除尘器　Cottrell electrostatic precipitator
科瓦铁镍钴合金　Kovar
颗粒物质　particulate matter
颗粒直径　particle diameter
壳层电子　shell electron
壳(层)模型　shell model
壳层坍缩　collapse of shell
可编程只读存储器　PROM（programmable read-only memory）
可编址寄存器　addressable register
可变磁阻　variable reluctance
可变电感　variable inductance
可变电感器　variable inductor；variometer
可变电感线圈　variable inductor
可变电抗器　varactor

可变电容器 adjustable capacitor; variable condenser; variodenser
可变电阻器 rheostat; variable resistor
可变负荷 variable load
可变负载 variable load
可变光阑 iris
可变焦比 zoom ratio
可变焦距镜头 varifocal lens
可变滤波器 variable filter
可变滤光片 variable filter
可变耦合器 variocoupler
可变速度 variable speed; variable velocity
可擦编程只读存储器 EPROM (erasable programmable read-only memory)
可重复性 reproducibility
可动平衡 mobile equilibrium
可动缺陷 mobile defect
可分裂物质 fissile material
可复现性 reproducibility
可更新能源 renewable energy source
可几速度 probable velocity
可积系统 integrable system
可见度 visibility
可见光 visible light
可见光谱 visible spectrum
可见(光谱)区 visible range
可靠性 reliability
可控硅整流器 silicon controlled rectifier
可裂变核 fissile nucleus; fissionable nucleus
可裂变物质 fissile material
可裂变性 fissility
可逆过程 reversible process
可逆机 reversible engine
可逆循环 reversible cycle
可溶性 solubility
可调电容器 adjustable capacitor
可调谐激光器 tunable laser
可调轴承 adjustable bearing
可调自耦变压器 variable autotransformer
可听度 audibility
可寻址存储器 addressable memory
可寻址寄存器 addressable register
可压缩性 compressibility
克 g (gram)
克尔解 Kerr solution
克尔效应 Kerr effect
克卡 therm; gram calorie
克拉 carat; karat
克拉克电池 Clark cell
克拉默斯定理 Kramers theorem
克拉尼图形 Chladni figures
克拉珀龙-克劳修斯方程 Clapeyron-Clausius equation
克莱因-戈尔登方程 Klein-Gordon equation
克劳 clausius
克劳修斯-莫索提方程 Clausius-Mossotti equation
克洛德循环 Claude cycle
克努森流 Knudsen flow
克努森效应 Knudsen effect
刻度板 scale plate
刻度尺 scale
刻度值 scale value
客观量度 objective measurement
空盒气压表 aneroid barometer
空化 cavitation
空间 space
空间电荷 space charge
空间电荷区 space charge region
空间反射 space reflection
空间反演 space inversion
空间格点 space lattice
空间关联 space correlation
空间静电凝集 electrostatic agglomeration in space
空间曲线 space curve
空间探测器 space probe
空间真空(度) vacuum of space
空气 air
空气弹道学 aeroballistics
空气当量 air equivalent

空气电介质 air dielectric
空气(电介质)电容器 air dielectric capacitor
空气电离 air ionization
空气动力学 aerodynamics
空气流量计 airflow meter
空气流速计 airmeter
空气流体动力学 aerohydrodynamics
空气摩擦 air friction
空气生物学 aerobiology
空气声学 aeroacoustics
空气湿度 air moisture
空气调节器 air conditioner
空气污染 air pollution；airborne contamination
空气芯 air core
空气制动机 airbrake
空气阻力 air friction
空气阻尼 air damping
空腔谐振模 cavity mode
空速 airspeed
空速管 Pitot tube
空调 air conditioner
空心线圈 air core coil
空穴 cavity；hole
空穴态 hole state
空穴现象 cavitation
孔 aperture；hole
孔径 aperture
孔径光阑 aperture stop
孔径角 angular aperture；aperture angle；opening angle
孔径综合 aperture synthesis
孔阑 aperture stop
孔特管 Kundt's tube
空格点 vacancy
空位 vacancy
空位群 vacancy cluster
空位团 vacancy cluster
控制棒 control rod
控制部件 CU（control unit）
控制参量 control parameter
控制杆 control rod
控制器 CU（control unit）
控制栅极 control grid
库仑 C（coulomb）
库仑常数 Coulomb constant
库仑场 Coulomb field
库仑定律 Coulomb's law
库仑规范 Coulomb gauge
库仑激发 Coulomb excitation
库仑计 coulometer；voltameter
库仑力 Coulomb force
库仑能 Coulomb energy
库仑碰撞 Coulomb collision
库仑势垒 Coulomb barrier
库珀不稳定性 Cooper instability
库珀对 Cooper pairs
库珀最小 Cooper minimum
夸克 quark
夸克禁闭 quark confinement
夸脱 qt（quart）
跨导 transconductance
跨声速的 transonic
快变量 fast variable
快度 rapidity
快堆 fast reactor
快过程 fast process
快速傅里叶变换 FFT（fast Fourier transform）
快中子 fast neutron
快中子增殖反应堆 fast breeder reactor
快子 tachyon
宽带 broadband；wideband
宽频带 wideband
宽谱带 broadband
矿灰 calx
扩程器 multiplier
扩散 diffusion
扩散泵 diffusion pump
扩散长度 diffusion length
扩散核函 diffusion kernel
扩散率 diffusivity
扩散能力 diffusivity
扩散器 diffuser
扩散系数 coefficient of diffusion
扩散云室 diffusion cloud chamber
扩散置限聚集 DLA（diffusion

limited aggregation)
扩束器 beam expander
扩压器 diffuser
扩音器 amplifier；loudspeaker；microphone
扩展 expansion
扩展 X 射线吸收精细结构 EXAFS (extended X-ray absorption fine structure)
扩展态 extended state
扩展性 expansivity

L

拉德　rad
拉格朗日方程　Lagrangian equation
拉格朗日函数　Lagrangian function
拉力　tension；traction force
拉曼(光)谱　Raman spectrum
拉曼激光器　Raman laser
拉曼散射　Raman scattering
拉曼效应　Raman effect
拉莫尔半径　Larmor radius
拉莫尔进动　Larmor precession
拉莫尔频率　Larmor frequency
拉莫尔旋进　Larmor precession
拉姆斯登目镜　Ramsden eyepiece
拉普拉斯方程　Laplace equation
拉伸模量　tensile modulus
拉伸弹簧　tension spring
拉伸应变　tensile strain
拉伸指数　tensile figure
拉乌耳定律　Raoult's law
拉应力　tensile stress
喇叭　loudspeaker
来斯利立方体　Leslie's cube
莱顿瓶　Leyden jar
莱曼系　Lyman series
兰金循环　Rankine cycle
兰姆凹陷　Lamb dip
兰姆移位　Lamb shift
兰氏温标　Rankine temperature scale
蓝(色偏)移　blue shift
蓝相　blue phase
朗伯　lambert
朗伯定律　Lambert's laws
朗道能级　Landau level
朗道阻尼　Landau damping
朗德 g 因子　Landé g-factor
朗德间隔定则　Landé interval rule
朗缪尔波　Langmuir wave
朗缪尔频率　Langmuir frequency
朗缪尔探针　Langmuir probe
朗缪尔振荡　Langmuir oscillation
朗之万方程　Langevin equation
朗之万函数　Langevin function
浪涌电流　surge current
劳埃德镜　Lloyd's mirror
劳厄对称　Laue-symmetry
劳厄对称群　Laue-symmetry group
劳厄法　Laue method
劳厄衍射　Laue diffraction
老花眼　presbyopia
老化　aging
勒　lx (lux)
勒克朗谢电池　Leclanché cell
勒克斯　lx (lux)
勒克斯计　luxmeter
勒让德变换　Legendre transformation
雷暴　thunderstorm
雷达　radar (radio detection and ranging)
雷姆　rem (roentgen equivalent man)
雷诺数　Reynolds number
镭　Ra (radium)
类磁通　fluxoid
类木行星　Jovian planet
类氢原子　hydrogen-like atom
类时　timelike
类时间隔　timelike interval

类时矢量 timelike vector
类时事件 timelike event
类星射电源 QSS (quasi-stellar radio source)
类星体 QSO (quasi-stellar object); quasar
类星星系 QSG (quasi-stellar galaxy)
棱镜 prism
棱镜光谱 prismatic spectrum
棱镜摄谱仪 prism spectrograph
棱锥体 pyramid
楞次定律 Lenz's law
冷槽 cold trap
冷等离子体 cold plasma
冷冻 refrigeration
冷冻干燥 freeze drying
冷冻混合物 freezing mixture
冷冻剂 coolant; refrigerant
冷发射 cold emission
冷光 cold light
冷阱 cold trap
冷聚变 cold fusion
冷凝 condensation; condensing
冷凝泵 condensation pump
冷凝抽机 condensation pump
冷凝剂 freezing mixture
冷凝器 condenser
冷凝吸附 cryosorption
冷却剂 coolant
冷源 heat sink
冷中子 cold neutron
冷子管 cryotron
厘米 cm (centimetre)
厘米·克·秒单位 c.g.s. units
厘米·克·秒制 c.g.s. system
厘米烛光 phot
离焦像 defocused image; out-of-focus image
离解 dissociation
离解热 heat of dissociation
离散本征值 discrete eigenvalue
离散流体模型 discrete fluid model
离散谱 discrete spectrum
离散图像 discrete picture
离心 centrifugation
离心泵 centrifugal pump
离心抽机 centrifugal pump
离心分离 centrifugation
离心(分离)机 centrifuge
离心加速度 centrifugal acceleration
离心力 centrifugal force
离心率 eccentricity
离心势垒 centrifugal barrier
离子 ion
离子半径 ionic radius
离子泵 ion pump
离子对 ion pair
离子发动机 ion engine
离子规 ionization gauge
离子轰击 ion bombardment
离子化 ionization
离子键 ionic bond
离子键合 ionic bonding
离子交换 ion exchange
离子交换吸附 ion exchange adsorption
离子晶体 ionic crystal
离子偶 ion pair
离子强度 ionic strength
离子散射谱学 ion scattering spectroscopy
离子束 ion beam
离子束发生器 ion-beam generator
离子束分析 IBA (ion-beam analysis)
离子微探针分析 ion-microprobe analysis
离子植入 ion implantation
离子注入 ion implantation
李(政道)模型 Lee model
里德伯常量 Rydberg constant
里德伯常数 Rydberg constant
里德伯分子 Rydberg molecule
里德伯(谱线)系 Rydberg series
里德伯原子 Rydberg atom
里克特震级 Richter magnitude
里氏震级 Richter magnitude
理查孙-德西曼方程 Richardson-

Dushman equation
理查孙方程 Richardson equation
理论 theory
理论力学 theoretical mechanics
理论强度 theoretical strength
理论物理学 theoretical physics
理论误差 theoretical error
理想成像 perfect imaging
理想导体 perfect conductor
理想晶体 ideal crystal
理想流体 ideal fluid
理想气体 ideal gas; perfect gas
理想气体温标 ideal gas thermometric scale
理想溶液 ideal solution; perfect solution
理想实验 gedanken experiment
理想透镜 perfect lens
理想约束 ideal constraint
锂电池 lithium battery
力 force
力倍增器 force multiplier
力比 force ratio
力臂 arm of force
力场 (force) field
力的分解 resolution of force
力的合成 composition of forces
力的平衡 equilibrium of forces
力的平行四边形 parallelogram of forces
力多边形 polygon of forces
力矩 moment (of force)
力螺旋 force screw
力偶 couple
力偶臂 arm of couple
力偶矩 moment of couple
力热效应 mechanocaloric effect
力线 lines of force
力心 centre of force
力学 mechanics
力学量 dynamical variable
力学品质因数 mechanical quality factor
力学平衡条件 mechanical equilibrium condition
力学系(统) mechanical system
力学运动 mechanical motion
力学质量 mechanical mass
历书 ephemeris
历书时 ET (ephemeris time)
立波 standing wave
立方的 isometric
立方晶体 cubic crystal
立方晶系 cubic system; isometric system
立方棱镜 block prism
立方厘米 cc (cubic centimetre)
立方米 stere
立方密堆积 cubic close packing
立方系 isometric system
立体的 three-dimensional
立体电影 stereoscopic film
立体光学 stereoptics
立体弧度 sr (steradian)
立体角 solid angle
立体摄影光学 stereoptics
立体声 stereophonic sound
利萨如图形 Lissajous figures
励磁 excitation
励磁曲线 excitation curve
粒子 particle
粒子加速器 particle accelerator
粒子浓度 particle concentration
粒子谱 particle spectrum
粒子束 particle beam
粒子数布居反转 population inversion
粒子数反转 population inversion
粒子物理学 particle physics
粒子性 corpuscular property
粒子直径 particle diameter
连通键 connected bond
连续本征值 continuous eigenvalue
连续波 continuous wave
连续波激光器 continuous wave laser
连续函数 continuous function
连续激光 continuous laser
连续介质 continuous medium; continuum
连续谱 continuous spectrum

连续区 continuum
连续统(一体) continuum
连续相 continuous phase
连续相变 continuous phase transition
连续性方程 continuity equation
敛集率 packing fraction
链式反应 chain reaction
链状结构 chain structure
良导体 good conductor
量程 range
量度 measurement
量能器 calorimeter
量热器 calorimeter
梁 beam
两次曝光全息术 double-exposure holography
两体碰撞 binary collision
两性离子 ampholyte ion; zwitterion
亮度 brightness; luminance; luminosity
亮度计 stilbmeter
亮视场 bright field
量纲 dimension
量纲分析 dimensional analysis
量级 magnitude
量子 quantum
量子产额 quantum yield
量子场论 quantum field theory
量子尺寸效应 quantum size effect
量子点 quantum dot
量子电动力学 QED (quantum electrodynamics)
量子电子学 quantum electronics
量子光学 quantum optics
量子化 quantization
量子化学 quantum chemistry
量子混沌 quantum chaos
量子霍尔效应 quantum Hall effect
量子阱 quantum well
量子理论 quantum theory
量子力学 quantum mechanics
量子模拟 quantum simulation
量子色动力学 QCD (quantum chromodynamics)
量子数 quantum number
量子数亏损 quantum defect
量子态 quantum state
量子统计法 quantum statistics
量子围栏 quantum corral
量子味动力学 QFD (quantum flavour dynamics)
量子物理学 quantum physics
量子效应 quantum effect
量子修正 quantum correction
量子引力 quantum gravity
量子引力动力学 QGD (quantum gravitational dynamics)
量子宇宙学 quantum cosmology
量子跃迁 quantum transition
列 train
列氏温标 Réaumur temperature scale
列线图 nomogram
裂变 fission
裂变参量 fission parameter
裂变产额 fission yield
裂变产物 fission product
裂变弹 fission bomb
裂变堆 fission reactor
裂变截面 fission cross-section
裂变径迹定年法 fission-track dating
裂变能 fission energy
裂变势垒 fission barrier
裂变碎片 fission fragment
裂变位垒 fission barrier
裂变阈 fission threshold
裂变中子 fission neutron
裂缝 slit
裂片反应 fragmentation reaction
裂纹扩展 crack propagation
林德循环 Linde cycle
林德液化机 Linde liquefier
临界参量 critical parameter
临界长度 critical length
临界点 critical point
临界反应 critical reaction

临界回授 critical reaction
临界角 critical angle
临界面 critical surface
临界速度 critical velocity
临界态 critical state
临界体积 critical volume
临界温度 critical temperature
临界现象 critical phenomenon
临界压力 critical pressure
临界压强 critical pressure
临界质量 critical mass
临界阻尼 critical damping
磷光 phosphorescence
磷光体 phosphor
磷青铜 phosphor bronze
灵敏电流计 galvanometer
灵敏度 sensitivity
菱面体格 rhombohedral lattice
菱形点阵 rhombohedral lattice
零点校正器 zero adjuster
零点能 zero-point energy
零点调整 zero adjustment
零读数 zero reading
零刻度 zero graduation
零(模)矢 null vector
零拍 zero beat
零位法 null method
流 current
流变学 rheology
流电(学) voltaism
流度 fluidity
流化 fluidization
流控技术 fluidics
流控学 fluidics
流量计 flowmeter
流量密度 flux density
流密度矢量 flux-density vector
流明 lm (lumen)
流体 fluid
流体动力学 fluid dynamics; hydrodynamics
流体静力学 hydrostatics
流体静压力 hydrostatic pressure
流体静压强 hydrostatic pressure
流体力学 fluid mechanics; hydromechanics
流体性 fluidity
流体压强计 manometer
流线 streamline
流线流 streamline flow
流线型 streamline
流星 meteor
流星体 meteoroid
流值 rhe
流质 fluid
琉态转变 glass transition
硫化镉激光器 CdS laser
六分仪 sextant
六角晶系 hexagonal system
六角密堆积 hexagonal close packing
漏磁通 leakage flux
漏电 electric leakage
漏电阻 leakage resistance
漏极 drain
漏气 vacuum leak
漏通量 leakage flux
漏泄电流 leakage current
漏泄因子 leakage factor
卢瑟福背散射 Rutherford backscattering
卢瑟福-玻尔原子模型 Rutherford-Bohr atom model
卢瑟福后向散射 Rutherford backscattering
卢瑟福散射 Rutherford scattering
卢瑟福(散射)实验 Rutherford experiment
卤素 halogen
录像带 videotape
路程 path
路径 path
路径积分 path integral
露 dew
露点 dew point
露点湿度计 dew-point hygrometer
铝硅镁合金 alsimag
铝镍钴合金 alnico
铝热剂 thermite
律音 note

滤泵 filter pump
滤波 filtration
滤波电容器 smoothing capacitor
滤波器 filter
滤出液 filtrate
滤光片 filter
滤色片 colour filter
滤色器 colour filter
孪激光器 twin laser
孪晶 twin crystal
孪晶带 twin band
孪晶体 bicrystal
孪晶轴 twinning axis
孪片磁体 twin magnets
孪三极管 twin triode
孪(生)像 twin image
孪生轴 twinning axis
乱真子 spurion
掠入射 glancing incidence
掠射角 glancing angle
伦琴 roentgen
轮 wheel
轮轴 wheel and axle
轮转线 cycloid
罗兰环 Rowland ring
罗盘 (magnetic) compass
罗斯合金 Rose's metal
罗素-桑德斯耦合 L-S coupling; Russell-Saunders coupling
罗谢尔盐 Rochelle salt
罗雄棱镜 Rochon prism
逻辑代数 algebra of logic
逻辑电路 logic circuit
螺钉 screw
螺杆 screw
螺距 pitch
螺绕环 torus
螺纹 screw thread
螺线 spiral
螺线管 solenoid
螺旋 screw
螺旋测径器 micrometer caliper
螺旋测微器 micrometer caliper
螺旋磁性 helimagnetism
螺旋度 helicity
螺旋法则 corkscrew rule
螺旋线 spiral
螺旋星系 spiral galaxy
螺旋性 helicity
螺状相 cholesteric phase
裸电荷 bare charge
裸核子 nucleor
裸奇异性 naked singularity
裸线 bare wire
洛伦兹变换 Lorentz transformations
洛伦兹力 Lorentz force
洛伦兹-洛伦茨方程 Lorentz-Lorenz equation
洛伦兹群 Lorentz group
洛施密特常量 Loschmidt number; Loschmidt's constant
落后 lag

Mm

马达 motor
马顿斯体 martensite
马尔可夫过程 Markov(ian) process
马尔特十字接头 Maltese cross
马格努斯效应 Magnus effect
马赫表 Machmeter
马赫计 Machmeter
马赫数 Mach number
马赫原理 Mach's principle
马克苏托夫望远镜 Maksutov telescope
马力 hp (horsepower)
马吕斯定律 Malus law
马略奥特定律 Mariotte's law
马氏体 martensite
码 yard
迈克耳逊测星干涉仪 Michelson stellar interferometer
迈克耳逊干涉仪 Michelson interferometer
迈克耳逊-莫雷实验 Michelson-Morley experiment
迈斯纳效应 Meissner effect
迈耶函数 Mayer function
麦 maxwell
麦克风 microphone
麦克劳德压力计 McLeod gauge
麦克劳德真空规 McLeod vacuum gauge
麦克劳林级数 Maclaurin's series
麦克斯韦 maxwell
麦克斯韦-玻尔兹曼分布 Maxwell-Boltzmann distribution
麦克斯韦-玻尔兹曼统计法 Maxwell-Boltzmann statistics
麦克斯韦方程组 Maxwell's equations
麦克斯韦关系 Maxwell relation
麦克斯韦速度分布 Maxwell velocity distribution
麦克斯韦速率分布 Maxwell speed distribution
麦克斯韦妖 Maxwell's demon
麦哲伦云 Magellanic clouds
脉冲 impulse
脉冲编码调制 PCM (pulse-code modulation)
脉冲持续时间 pulse duration; pulse width
脉冲触发 pulse triggering
脉冲传输 pulse transmission
脉冲电压 impulse voltage
脉冲发生器 pulser
脉冲反射 pulse reflection
脉冲激光 spike
脉冲激光器 pulsed laser
脉冲间隔 pulse spacing
脉冲宽度 pulse duration; pulse width
脉冲雷达 pulse radar
脉冲谱 pulse spectrum
脉冲强度 impulse strength; pulse strength
脉冲调制 pulse modulation
脉冲响应 impulse response
脉冲星 pulsar
脉冲修尖电路 peaking circuit
脉冲展宽 pulse stretching
脉动 pulsation
脉动式喷气发动机 pulse jet
满带 filled band
漫反射 diffuse reflection; scattered reflection

漫射 diffusion
漫射光 diffused light
漫射体 diffuser
慢中子 slow neutron
盲点 blind spot
毛发湿度计 hair hygrometer
毛细管 capillary tube
毛细管孔 capillary bore
毛细现象 capillarity
毛细作用 capillarity; capillary action
锚环 anchor ring
媒质 medium
煤 coal
美国国家标准协会 ANSI (American National Standards Institute)
美国信息交换标准码 ASCII (American Standard Code for Information Interchange)
镁铝合金 Magnalium
镁诺克斯合金 Magnox
门(电路) gate
钔 Md (mendelevium)
蒙次合金 Muntz metal
蒙次黄铜 Muntz metal
蒙特卡罗法 Monte Carlo method
锰镍铜合金 manganin
锰铜 manganin
弥散 dispersion; dissipation
弥散相 dispersed phase
醚 ether
米 m (metre)
米·千克·秒制 m. k. s. units
米尺 metre rule
米勒指数 Miller indices
米制 metric system
密度 density
密度波动 density fluctuation
密度泛函法 density functional method
密度计 densitometer
密度矩阵 density matrix
密度涨落 density fluctuation
密堆积 close packing
密封舱 pressurized cabin
密封环 packing ring
密立根油滴实验 Millikan oil-drop experiment
幂 power
幂次 degree
幂函数 power function
面包板 breadboard
面电流 surface current
面电流密度 surface current density
面积分 surface integral
面积仪 planimeter
面结型晶体管 junction transistor
面缺陷 planar defect
面心格 face-centred lattice
面心立方体 f. c. c. (face-centred cubic)
瞄准器 finder
瞄准线 line of sight
秒 s (second)
秒差距 parsec
闵可夫斯基几何 Minkowski geometry
闵可夫斯基空间 Minkowski space
敏感元件 sensor
明暗界限 terminator
明(视)场 bright field
明视距离 distance of distinct vision
明线 bright line
模 mode
模糊图像 blurred image
模结构 mode configuration
模量 modulus
模拟 simulation
模拟板 breadboard
模拟电路 analogue circuit
模拟加法器 analogue adder
模拟信号 analogue signal
模耦合 mode coupling
模牵引效应 mode pulling effect
模式锁定 mode-locking
模数 mode number; modulus

模数转换 analogue-to-digital conversion
模数转换器 ADC (analogue-to-digital converter)
模态 modality
模图样 mode pattern
模推斥效应 mode pushing effect
模型 model
模序列 mode sequence
模抑制 mode suppression
膜沸腾 film boiling
膜片 membrane
摩擦 friction
摩擦电 triboelectricity
摩擦发光 triboluminescence
摩擦力 friction (force)
摩擦起电 charging by friction; electrification by friction; triboelectricity; triboelectrification
摩擦系数 coefficient of friction
摩擦学 tribology
摩尔 mole
摩尔导电率 molar conductivity
摩尔的 molar
摩尔分数 mole fraction
摩尔热容 molar heat capacity
摩尔体积 molar volume; molecular volume
磨蚀剂 abrasive
魔(鬼楼)梯 devil's staircase
末速(度) final velocity
末态 final state
莫尔斯电报机 Morse recorder
莫尔斯电码 Morse alphabet
莫尔条纹 moiré fringe
莫霍罗维奇界面 Mohorovičić discontinuity
莫塞莱定律 Moseley's law
莫氏硬度标 Moh hardness scale
墨菲公式 Murphy formula
母核 parent nucleus
母结构 parent structure
母体 parent
母线 bus; generatrix
母液 mother liquor; mother solution
姆欧 mho
目标 object; target
目标强度 target strength
目测光度计 visual photometer
目镜 eyepiece; ocular
目镜分光镜 eyepiece spectroscope
目视双星 visual binary
穆斯堡尔效应 Mössbauer effect

Nn

镎 Np（neptunium）
镎系 neptunium series
纳米 nm（nanometre）
纳米技术 nanotechnology
纳秒 ns（nanosecond）
纳皮尔对数 Napierian logarithm
纳维-斯托克斯方程 Navier-Stokes equation
钠灯 sodium lamp
钠汽灯 sodium-vapour lamp
氖灯 neon lamp
奈尔温度 Néel temperature
奈培 neper
耐火材料 refractory
耐火的 refractory
耐熔质 refractory
南极光 aurora australis
难磁化方向 hard direction for magnetization
难磁化轴 hard axis of magnetization
难熔的 refractory
挠度 deflection
挠矩 bending moment
挠率 torsion
挠率张量 torsion tensor
挠曲线 twisted curve
挠胁变 bending strain
挠胁强 bending stress
脑电描记法 electroencephalography
脑电图 EEG（electroencephalogram）
脑电图学 electroencephalography
内禀半导体 intrinsic semiconductor
内禀方程 intrinsic equation
内禀角动量 intrinsic angular momentum
内禀载流子 intrinsic carrier
内部变换 internal conversion
内部对称性 internal symmetry
内插 interpolation
内插多项式 interpolation polynomial
内电阻 internal resistance
内反射 internal reflection
内光电效应 internal photoelectric effect
内耗 internal friction
内积 scalar product
内建场 built-in field
内聚力 cohesion；cohesive force
内聚能 cohesive energy
内窥镜 endoscope；introscope
内力 internal force
内摩擦 internal friction
内能 internal energy；intrinsic energy
内燃机 internal combustion engine
内液增阻效应 Hall effect
内转换 internal conversion
内转换系数 internal conversion coefficient
内阻 internal resistance
能包 packet of energy
能层 ergosphere
能带 （energy）band
能带变窄 band narrowing
能带结构 energy-band structure
能带理论 band theory
能带论 （energy-）band theory

能级 (energy) level
能级分裂 split of energy level
能级交叉效应 level-crossing effect
能级密度 level density
能级图 energy level diagram
能见度 visibility
能量 energy
能量不灭 conservation of energy；energy conservation
能量均分 equipartition of energy
能量均分定理 equipartition theorem
能量利用 energy use
能量密度 energy density
能量守恒 conservation of energy；energy conservation
能量守恒定律 law of conservation of energy
能量守恒与转化定律 law of conservation of energy
能量损失 energy loss
能流 (energy) flux
能流密度 energy flux density
能谱 energy spectrum
能斯特定理 Nernst theorem
能斯特效应 Nernst effect
能通量 energy flux
能隙 energy gap
能源消费 energy use
尼科耳棱镜 Nicol prism
尼特 nit
霓 secondary rainbow
霓虹灯 neon lamp
拟晶体 crystalloid
逆操作 inverse operation
逆磁光效应 inverse magnetooptical effect
逆磁性 diamagnetism
逆反应 reverse reaction
逆函数 inverse function
逆康普顿散射 inverse Compton scattering
逆康普顿效应 inverse Compton effect
逆渗透 reverse osmosis
逆弹性 dielasticity
逆推力 reverse thrust
逆温层 inversion layer
逆相 antiphase；opposite phase
逆相位 antiphase
逆相序阻抗 negative sequence impedance
逆行 retrograde motion
年 year
黏度计 viscometer
黏度指数 viscosity index
黏附 adhesion
黏附力 cohesive force；stick force
黏附系数 sticking coefficient
黏固 cementation
黏弹性 viscoelasticity
黏性 viscosity
黏性流体 viscous fluid
黏性系数 coefficient of viscosity
黏滞度 viscosity
黏滞流体 viscous fluid
黏滞摩擦 viscous friction
黏滞系数 coefficient of viscosity
黏滞性 viscosity
镍镉电池 nickel-cadmium cell
镍铬合金 Nichrome
镍铬恒弹性钢 Elinvar
镍铝硅锰(电偶)合金 alumel
镍铁电池 nickel-iron battery；NiFe battery
凝点 freezing point
凝固 freezing；solidification
凝固点 freezing point；solidifying point
凝固点降低 depression of freezing point
凝固温度 freezing point
凝胶 gel
凝胶化 gelation
凝胶化点 gelation point
凝胶模型 jellium model
凝胶相 gel phase
凝结 coagulation；condensation；condensing

凝晶质 crystalloid
凝聚态物理学 condensed-matter physics
凝聚体 condensed matter
凝聚体物理学 condensed-matter physics
凝团 clump
牛顿 N（newton）
牛顿第二定律 Newton second law
牛顿第三定律 Newton third law
牛顿第一定律 Newton first law
牛顿公式 Newton's formula
牛顿环 Newton's rings
牛顿冷却定律 Newton's law of cooling
牛顿力学 classical mechanics；Newtonian mechanics
牛顿流体 Newtonian fluid
牛顿色盘 Newton's disk
牛顿条纹 Newton's rings
牛顿望远镜 Newtonian telescope
牛顿引力定律 Newton's law of gravitation
牛顿运动三定律 Newton's laws of motion
扭秤 torsion balance
扭角仪 troptometer
扭矩 torque；twisting moment
扭力 twisting force
扭力计 troptometer
扭量 twistor
扭率张量 torsion tensor
扭曲 twist
扭胁强 torsional stress；twisting stress
扭应力 torsional stress；twisting stress
扭转 torsion；twist
扭转刚度 torsional rigidity
扭转晶界 twist boundary
扭转模 torsional mode
扭转仪 troptometer
扭转运动 twisting motion
扭转振动 torsional vibration
浓差电池 concentration cell
浓缩 enrichment
诺埃曼（互感）公式 Neumann formula
诺克斯 nox
诺模图 nomogram
诺特定理 Noether's theorem
锘 No（nobelium）

Oo

欧 Ω（ohm）
欧几里得维数 Euclidean dimension
欧拉方程 Euler equation
欧拉角 Eulerian angle
欧拉流体动力学方程 Euler equations for hydrodynamics
欧拉运动学方程 Eulerian kinematical equations
欧姆 Ω（ohm）
欧姆表 ohmmeter
欧姆电阻 ohmic resistance
欧姆定律 Ohm's law
欧姆加热 ohmic heating
欧姆接触 ohmic contact
欧姆损耗 ohmic loss
欧洲粒子物理研究所 CERN（Conseil Européen pour la Recherche Nucléaire）
偶分布函数 pair distribution function
偶奇核 even-odd nucleus
偶极 dipole
偶极辐射 dipole radiation
偶极近似 dipole approximation
偶极矩 dipole moment
偶极-偶极相互作用 dipole-dipole interaction
偶极子 dipole；doublet
偶偶核 even-even nucleus
耦合 coupling
耦合摆 coupled pendulums
耦合不足 undercoupling
耦合常数 coupling constant
耦合电路 coupled circuit
耦合方式 coupling scheme
耦合(共振)腔 coupled resonators
耦合系数 coupling coefficient

Pp

爬膜效应　creep film effect
帕　pa（pascal）
帕斯卡定律　Pascal's law
帕斯卡　Pa（pascal）
帕邢-巴克效应　Paschen-Back effect
帕邢系　Paschen series
拍　beats
拍频　beat frequency
排列　permutation
排气(装置)　exhaust
排他原理　exclusion principle
排液装置　drain
盘管　coil
庞加莱球　Poincaré sphere
庞加莱群　Poincaré group
旁路　by-pass
旁频带　sideband
抛光　burnishing；politure
抛体　projectile
抛体运动　projectile motion
抛物面　paraboloid
抛物面反射镜　parabolic reflector
抛物面反射器　paraboloidal reflector
抛物面镜　paraboloidal mirror
抛物线　parabola；paracurve
泡利不相容原理　Pauli exclusion principle
泡沫浮选　froth flotation
泡沫室　bubble chamber
炮铜　gun metal
佩尔捷效应　Peltier effect
配电盘　distributor
配分　partition
配分函数　partition function
配分系数　partition coefficient
配体场论　ligand field theory
配位数　coordination number
喷流　jet stream
喷气　jet
喷气发动机　jet engine
喷气式飞机　jet plane
喷气式螺旋桨　jet propeller
喷气推进　jet propulsion
喷气引擎　ramjet
喷射器　jet
喷射推进器　jet propeller
喷注　jet
喷注泵　jet pump
喷注抽机　jet pump
喷注簇射　jet shower
喷注模型　jet model
喷注起电　jet electrification
喷嘴　jet
硼电离室　boron chamber
硼计数管　boron counter tube
膨胀　dilatancy；dilatation；dilation；expansion；inflation
膨胀计　dilatometer
膨胀率　expansivity
膨胀系数　coefficient of expansion
碰撞　collision
碰撞参量　impact parameter
碰撞参数　impact parameter
碰撞弛豫　collision relaxation
碰撞电离　impact ionization；ionization by collision
碰撞过程　collision process
碰撞加热　collisional heating
碰撞截面　collision cross-section
碰撞密度　collision density

碰撞频率 collision frequency
碰撞起电 collision electrification
碰撞强度 collision strength
碰撞时间 collision time
碰撞守恒量 collisionally conserved quantity
碰撞增宽 collision broadening; impact broadening
碰撞展宽 collision broadening; impact broadening
劈理 cleavage
皮拉尼压力计 Pirani gauge
皮拉尼真空规 Pirani gauge
皮拉尼真空计 Pirani gauge
皮托管 Pitot tube
疲乏 fatigue
疲乏极限 fatigue limit
疲劳 fatigue
疲劳限度 fatigue limit
匹配滤波器 matched filter
偏差 declination; deflection; deviation
偏差校正 offset correction
偏差数 deviation
偏导数 partial derivative
偏光镜 polariscope
偏角 (angle of) declination
偏食 partial eclipse
偏微分方程 partial differential equation
偏微商 partial derivative
偏向角 angle of deviation
偏斜 deflection
偏斜角 offset angle
偏心曲柄机构 offset crank mechanism
偏心凸轮 offset cam
偏移 offset
偏移目镜 offset eyepiece
偏移射束 offset beam
偏移误差 offset error
偏振 polarization
偏振度 degree of polarization
偏振方向 polarization direction
偏振光镜 polariscope
偏振计 polarimeter
偏振角 angle of polarization
偏振面 plane of polarization
偏振片 polaroid
偏振椭圆 polarization ellipse
偏振效应 polarization effect
偏振元件 polarizer
偏置 offset
偏置法 offsetting
偏轴角 off-axis angle
偏转 deflection; deviation
偏转函数 deflection function
偏转后加速 post-deflection acceleration
偏转线圈 deflecting coil
漂浮 floating
漂移 deviation
漂移迁移率 drift mobility
漂移室 drift chamber
漂移速度 drift velocity
漂白 decoloration
频带 band
频率 frequency
频率计 frequency meter
频率匹配 frequency matching
频率牵引 frequency pulling
频率调制 FM (frequency modulation)
频率推斥 frequency pushing
频谱 frequency spectrum
频谱分解 spectral decomposition
频谱分析 spectral analysis
频谱密度 spectral density
频谱面 frequency plane
频闪测速计 strobotach
频闪(观测)仪 stroboscope
频移 frequency shift
频域 frequency domain
品质因数 factor of quality; Q factor (quality factor)
平凹透镜 plano-concave lens
平动 translation
平动加速度 translational acceleration
平动能 translational energy
平动速度 translational velocity

平动态 translational state
平动自由度 translational degree of freedom
平方反比定律 inverse-square law
平方根 square root
平衡 balance; equilibrium; poise
平衡电桥 balanced bridge
平衡方程 balance equation
平衡态 equilibrium state
平衡条件 equilibrium condition
平衡位形 equilibrium configuration
平衡位置 equilibrium position
平衡形变 equilibrium deformation
平滑电容器 smoothing capacitor
平截头台 frustum
平近点角 mean anomaly
平晶 optical flat
平均的 average; mean
平均偏差 mean deviation
平均射程 mean range
平均寿命 mean lifetime
平均数 average; mean
平均速度 mean velocity
平均速率 average speed; mean speed
平均误差 average error; mean error
平均自由程 mean free path
平均自由时间 mean free time
平流 advection
平流层 advective layer; stratosphere
平炉 open-hearth furnace
平面 plane
平面波 plane wave
平面波玻恩近似 plane wave Born approximation
平面二极管 planar diode
平面干涉仪 flat interferometer
平面工艺 planar technology
平面光栅 plane grating; two-dimensional grating
平面结 planar junction
平面晶体管 planar transistor
平面镜 plane mirror
平面偏振 plane polarization
平面偏振光 plane polarized light
平面全息图 plane hologram
平面阵列 planar array
平太阳日 mean solar day
平凸透镜 plano-convex lens
平行光管 collimator
平行光束 parallel beam
平行力 parallel force
平行流 parallel flow
平行六面体 parallelepiped
平行射线束 parallel beam
平行轴定理 parallel axis theorem
平行自旋 parallel spins
平移 translation
平移对称 translational symmetry
平移对称素 translational symmetry element
平移对称元 translational symmetry element
平移加速度 translational acceleration
平移运动 translational motion
平印照相制版 photolithography
平直时空 flat spacetime
坪区 plateau (region)
坪区扩散 plateau diffusion
屏 screen
屏蔽 mask; shielding
屏蔽核素 shielded nuclide
屏蔽效应 shielding effect
屏幕 screen
屏栅 shield grid
屏栅极 screen grid
屏像 screen image
钋 Po (polonium)
坡莫合金 permalloy
坡印亭矢量 Poynting vector
破坏对称性 broken symmetry
破坏性干扰 destructive interference

镤 Pa（protactinium）
普丰德级数 Pfund series
普丰德系 Pfund series
普朗克长度 Planck length
普朗克常量 Planck constant
普朗克单位 Planck units
普朗克辐射定律 Planck's radiation law
普朗克公式 Planck formula
普朗克假说 Planck's hypothesis
普朗克质量 Planck mass
普雷伏热交换理论 Prevost's theory of heat exchange
普适常量 universal constant
普适恒量 universal constant
普适气体常量 universal gas constant
普适性 universality
普通物理学 general physics
谱 spectrum
谱带 spectral band
谱函数 spectral function
谱类 spectral class
谱线 spectral line
谱线宽度 line breadth；line width
谱线轮廓 line profile
谱线强度 line strength
谱线系 line series；spectral series
谱线形状 line shape
谱线引力位移 Einstein shift
谱线增宽 line broadening
谱型 spectral class

Qq

期待值 expectation value
期望值 expectation value
齐明镜 aplanat
齐明透镜 aplanatic lens
齐纳二极管 Zener diode
齐纳击穿 Zener breakdown
奇点 singularity
奇怪吸引子 strange attractor
奇特分子 exotic molecule
奇特态 exotic state
奇特原子 exotic atom
奇异粒子 strange particle
奇异扰动 singular perturbation
奇异性 singularity；strangeness
起磁 magnetization
起点读数 zero reading
起电 electrification；electrization
起电盘 electrophorus
起伏 fluctuation
起偏角 polarizing angle
起偏器 polarizer
起偏(振)镜 polarizer
起偏振片 polarizing plate
气垫桌 air table
气动工作台 air table
气动弹性 aeroelasticity
气动弹性力学 aeroelasticity
气动阻力 aerodynamic drag
气辉 airglow
气辉层 airglow layer
气辉发射 airglow emission
气冷反应堆 gas-cooled reactor
气流 airflow；airstream
气密性 airtightness
气敏器件 gas sensor
气泡室 bubble chamber
气泡栓塞症 aeroembolism
气溶胶 aerosol
气塞 air plug
气体 gas
气体比重计 aerometer
气体常数 gas constant
气体定律 gas laws
气体动理论 kinetic theory of gases
气体动理学 gas kinetics
气体动力学 gas dynamics
气体分子运动论 gas kinetics；kinetic theory of gases
气体激光器 gas laser
气体敏感器 gas sensor
气体燃化计 eudiometer
气体温度计 gas thermometer
气体压力计 tensimeter
气雾剂 aerosol
气相色谱分析 gas chromatography
气相色谱学 gas chromatography
气象台 observatory
气压测高计 pressure altimeter
气压计 barograph；barometer；manometer
气液连续性 gas-liquid continuity
汽 vapour
汽点 steam point
汽化 vaporization
汽化潜热 latent heat of vaporization
汽化热 heat of vaporization
汽蚀 cavitation
汽相 vapour phase
汽相沉积法 vapour-deposition method

汽压 vapour tension
汽油机 gasoline engine
器件 device
千分测径规 micrometer caliper
千分尺 micrometer
千赫 kHz (kilohertz)
千斤顶 jack
千卡 Calorie; kilocalorie; kilogramme calorie
千克 kg (kilogramme)
千米 km (kilometre)
千瓦 kW (kilowatt)
千瓦时 kWh (kilowatt-hour)
千兆 billion
千兆电子伏特 BeV (billion-electronvolt); GeV (gigaelectronvolt)
迁移理论 transport theory
迁移率 (drift) mobility
迁移率边 mobility edge
迁移率隙 mobility gap
迁移平均自由程 transport mean free path
迁移数 transport number
迁移系数 transport coefficient
牵连惯性力 convected inertial force
牵连加速度 convected acceleration
牵连速度 convected velocity; velocity of following
牵连运动 convected motion
牵引力 traction force
牵引阻力 tractive resistance
铅铋锡易熔合金 Rose's metal
铅垂线 plumb line; vertical line
铅酸电池 lead-acid cell
铅酸蓄电池 lead-acid accumulator
前导波 leading wave
前加速器 preaccelerator
前进波 advancing wave; progressive wave
前向散射 forward scattering
前置放大 preamplification
前置放大器 preamplifier
前置棱镜 fore-prism
前子 preon
钱德拉塞卡极限 Chandrasekhar limit
钱德勒数 Chandler number
潜热 latent heat
潜望镜 periscope
潜像 latent image
浅结 shallow junction
浅能级 shallow level
欠校正透镜 undercorrected lens
欠耦合 undercoupling
欠阻尼 underdamping
腔 cavity
腔口速度 muzzle velocity
腔模 cavity mode
腔体谐振器 cavity resonator
腔振荡 cavity oscillation
腔子 caviton
强度 intensity
强度变量 intensive variable
强度级 volume level
强度量 intensive quantity
强耦合 strong coupling
强迫对流 forced convection
强迫振动 forced vibration
强湍动 strong turbulence
强相互作用 strong interaction
强制对流 forced convection
强制锁模 forced mode-locking
强子 hadron
强子动力学 hadrodynamics
强子化 hadronization
强子量能器 hadron calorimeter
强子型衰变 hadronic decay
乔吉制 Giorgi system
桥臂 arm of the bridge
桥键 bridge bond
桥接电路 bridge circuit
桥接原子 bridging atom
桥式电路 bridge circuit
桥式整流器 bridge rectifier
切连科夫辐射 Cerenkov radiation
切连科夫计数器 Cerenkov counter; Cerenkov detector

切连科夫探测器 Cerenkov counter；Cerenkov detector
切线 tangent
切向加速度 tangential acceleration
切向速度 tangential velocity
切向胁强 tangential stress
切向应力 tangential stress
切应变 shearing strain
切展线 involute
亲电体 electrophile
擒纵机构 escapement
青铜 bronze
轻核 light nucleus
轻子 lepton
轻子型衰变 leptonic decay
氢 H (hydrogen)
氢半电池 hydrogen half cell
氢弹 hydrogen bomb
氢电极 hydrogen electrode
氢光谱 hydrogen spectrum
氢弧灯 hydrogen arc lamp
氢键 hydrogen bond
氢离子 hydrogen ion
倾度 inclination
倾角 (angle of) dip；inclination；obliquity
倾腔激光器 cavity-dumped laser
倾斜 obliquity
倾斜度 dip；gradient
倾斜计 inclinometer
倾斜因数 inclination factor；obliquity factor
倾斜因子 inclination factor；obliquity factor
清洁处理 cleaning
清洗 cleaning
求积仪 planimeter
求像(作图)法 image construction
球差 spherical aberration
球极坐标 spherical polar coordinates
球径计 spherometer
球面波 spherical wave
球面度 sr (steradian)
球面镜 spherical mirror
球面镜共振腔 spherical mirror resonator
球面投影 spherical projection
球面透镜 spherical lens
球面像差 spherical aberration
球体 sphere
球透镜 globe lens
球谐函数 spherical harmonics
球形核 spherical nucleus
球状闪电 ball lightning
球状星团 globular cluster
区域精炼 zone refining
区域提纯 zone refining
曲率 curvature
曲率半径 radius of curvature
曲率中心 centre of curvature
曲线拟合 curve fitting
曲线图 graph
曲线运动 curvilinear motion
曲轴 crankshaft
驱动力 driving force
屈服 yield
屈服点 yield point
屈光度 dioptre
屈光学 dioptrics
趋肤深度 skin depth
趋肤效应 Kelvin effect；skin effect
取景器 finder
取向 orientation
取向附生 epitaxial growth；epitaxy
取样 sampling
取样器 sampler
去磁 degaussing；demagnetization
去磁因数 demagnetization factor
去活化 deactivation
去激发 deexcitation
去激活 deactivation
去激励 deexcitation
去焦 defocusing
去色 decoloration
去像散镜头 anastigmatic lens
全并行模数转换器 all parallel

ADC
全波带接收机 all-wave receiver
全波带振荡器 all-wave oscillator
全波段辐射计 panradiometer
全波片 full-wave plate
全波天线 all-band aerial
全波整流器 full-wave rectifier
全剥等离子体 stripped plasma
全串行模数转换器 all-serial ADC
全反射 total reflection
全反射角 angle of total reflection
全景 full view
全景全息图 full view hologram
全聚反光装置 holophote
全孔径 full aperture
全口径 full aperture
全内反射 total internal reflection
全能峰 full energy peak
全色图像 full-colour image
全食 total eclipse；totality
全视图 full view
全通滤波器 all-pass filter
全通滤光片 all-pass optical filter
全同粒子 identical particles
全微分 perfect differential
全息(摄影)术 holography
全息图 hologram
全向天线 omnidirectional aerial
缺级 missing order
缺位 vacancy
缺陷 defect
缺序 missing order
群 group
群集 cluster
群聚压缩 bunch compression
群速(度) group velocity

Rr

燃料　fuel
燃料电池　fuel cell
燃料元件　fuel element
燃气轮机　gas turbine
燃烧　combustion
燃烧热　heat of combustion
燃素　phlogiston
燃素学说　phlogiston theory
染料激光器　dye laser
染色性　chromaticity
扰动媒质　turbulent medium
绕法　winding
绕射　diffraction
绕射条纹　diffraction fringe
绕转　revolution
绕组　winding
热　heat
热泵　heat pump
热变像管　thermal image converter
热场　thermal field
热成像　thermal imaging
热传导　heat conduction
热传递　thermal transmission
热当量　thermal equivalent
热导计　katharometer
热导率　heat conductivity; temperature conductivity; thermal conductivity
热导体　thermal conductor
热导性　heat conductivity
热等离子体　hot plasma
热电　pyroelectricity; thermoelectricity
热电堆　thermopile
热电偶　thermocouple
热电偶安培计　thermoammeter
热电偶接头　thermojunction
热电偶温度计　thermocouple thermometer
热电偶真空计　thermocouple gauge
热电效应　pyroelectric effect; thermoelectric effect
热电学　pyroelectricity
热电子　thermion; thermoelectron
热电子发射　thermionic emission
热阀　thermal valve
热法拉　thermal farad
热反馈　thermal feedback
热反应堆　thermal reactor
热峰　thermal spike
热辐射　heat radiation; temperature radiation; thermal radiation
热辐射体　thermal radiator
热负荷　thermal load
热功当量　mechanical equivalent of heat; thermal equivalent of work
热光　thermal light
热函　enthalpy
热核弹　fusion bomb
热核等离子体　thermonuclear plasma
热核反应　thermonuclear reaction
热核反应堆　thermonuclear reactor
热核聚变　thermonuclear fusion
热核武器　thermonuclear weapon
热后效　thermal aftereffect
热化学　thermochemistry

热汇 heat sink; thermal anchoring
热机 heat engine
热机效率 efficiency of heat engine
热激发 thermal excitation
热激活 thermal activation
热激励 thermal excitation
热寂 heat death
热交换器 heat exchanger
热绝缘 heat insulation; thermal insulation
热绝缘体 thermal insulator
热开关 thermal switch
热库 heat reservoir
热扩散 thermal diffusion; thermodiffusion
热离子 thermion
热离子传导 thermionic conduction
热离子电流 thermionic current
热离子二极管 thermionic diode
热离子发射 thermionic emission
热离子放大器 thermionic amplifier
热离子管 thermionic tube; thermionic valve
热离子学 thermionics
热离子阴极 thermionic cathode
热离子整流器 thermionic rectifier
热离子转换器 thermionic converter
热力层结 thermal stratification
热力学 thermodynamics
热力学第二定律 second law of thermodynamics
热力学第零定律 zeroth law of thermodynamics
热力学第三定律 third law of thermodynamics
热力学第一定律 first law of thermodynamics
热力学平衡 thermodynamic equilibrium
热力学温标 thermodynamic scale
热力学温度 thermodynamic temperature
热力学因子 thermodynamic factor
热量 heat
热量单位 thermal unit
热量计 calorimeter
热裂(作用) thermal fragmentation
热敏电阻 thermal resistor; thermistor
热能 heat energy; thermal energy
热能化 thermalization
热膨胀 thermal expansion
热膨胀系数 thermal expansion coefficient
热疲劳 thermal fatigue
热平衡 heat balance; thermal equilibrium
热平衡条件 thermal equilibrium condition
热屏 heat shield
热扰动 thermal agitation
热容 heat capacity
热容量 heat capacity; thermal capacity
热骚动 thermal agitation
热素 phlogiston
热速度 thermal velocity
热态 thermal state
热通量 thermal flux
热透镜 thermal lens
热透镜效应 thermal lensing effect
热稳定性 thermal stability
热线式电气测试仪表 hot-wire instrument
热效率 thermal efficiency
热胁强 thermal stress
热学 thermology
热学分析 thermoanalysis
热压结 sintering
热阴极 thermionic cathode
热阴极电子管 thermionic valve

热应力 thermal stress
热源 heat reservoir；heat source
热运动 thermal motion
热涨落 thermal fluctuation
热振动 thermal vibration
热值 calorific value；thermal value
热质说 caloric theory of heat
热致发光 thermoluminescence
热致发光测定年代 thermoluminescent dating
热滞后 thermal hysteresis
热滞弹性 thermoanelasticity
热中子 thermal neutron
热阻 thermal resistance
热阻率 thermal resistivity
热阻系数 thermal resistivity
人存原理 anthropic principle
人工放射性 artificial radioactivity
人工智能 AI（artificial intelligence）
人机工程学 ergonomics
人为嬗变 artificial transmutation
人眼光学 ophthalmic optics
人造卫星 artificial satellite
刃 ket
任意子 anyon
韧性断裂 ductile fracture
轫 brake
轫致辐射 Bremsstrahlung
日 day；sun
日珥 solar prominence
日光灯 fluorescent lamp
日冕 corona
日偏食 partial solar eclipse
日食 solar eclipse
日心说 heliocentric theory
日心体系 heliocentric system
日照辐射 insolation
容抗 capacitive reactance
容量 capacity；volume
容量计 volumeter
容限 tolerance
容许带 allowed band
容许负载 allowable load
容许过载 permissible overload
容许偏差 allowable deviation；tolerance deviation
容许误差 admissible error；allowable error
容许应力 permissible stress
容许跃迁 allowed transition；permitted transition
容许转变 allowed transition
溶度 solubility
溶度积 solubility product
溶度限 solubility limit
溶度指数 solubility exponent
溶剂 solvent
溶胶 sol
溶解 deliquescence；solution
溶解热 heat of solution
溶解物 solute
溶解系数 solubility coefficient
溶线 solubility curve；solvus
溶液 solution
溶质 solute
溶致液晶 lyotropic liquid crystal
熔点 fusion point；m. p.（melting point）
熔断器 fuse
熔断器盒 fuse box
熔合反应 fusion reaction
熔化 fusion；melt
熔化分离作用 liquation
熔化潜热 latent heat of fusion
熔化热 heat of fusion；melting heat
熔化物 melt
熔剂 flux
熔接 fusion
熔解热 heat of fusion
熔炼 smelting
熔融 melt
熔融石 tektite
熔丝 fuse
熔体 melt
蠕变 creep
蠕动效应 reptation effect
乳胶 emulsion

乳状液 emulsion
入日射 incoming solar radiation
入射 incidence
入射波 incident wave
入射道 incoming channel
入射光瞳 entrance pupil
入射角 angle of incidence; incidence (angle)
入射粒子 incident particle
入射(平)面 plane of incidence
入射束 incoming beam
入射线 incident ray
入射轴 axis of incidence
入瞳 entrance pupil
软磁材料 soft magnetic material
软辐射 soft radiation
软件 software
软流层 asthenosphere
软流圈 asthenosphere
软模 soft mode
软盘 diskette; floppy disk
软钎料 solder
软水 soft water
软水剂 water softener
软铁 soft iron
瑞利-金斯公式 Rayleigh-Jeans formula
瑞利判据 Rayleigh criterion
瑞利散射 Rayleigh scattering
瑞利数 Rayleigh number
润滑(作用) lubrication
弱耦合 weak coupling
弱相互作用 weak interaction
弱阻尼 underdamping

Ss

塞曼分裂　Zeeman splitting
塞曼效应　Zeeman effect
三倍倍频器　trebler
三重积分　triple integral
三重态　triplet (state)
三重线　triplet
三重线态　triplet state
三次倍频器　triductor
三次方程　cubic equation
三方点格　rhombohedral lattice
三分子模型　trimolecular model
三合薄透镜　triplet thin lens
三合透镜　triplet (lens)
三合星　triple star
三级像差　third order aberration
三极管　triode
三角函数　trigonometric function
三角晶系　trigonal system
三角棱镜　triangular prism
三角系晶体　trigonal crystal
三角形波　triangular wave
三脚放大镜　tripod magnifier
三脚架　tripod
三面体棱镜　triangular prism
三面直角棱镜　corner cube
三色视觉　trichromatic vision
三态点　triple point
三体碰撞　triple collision
三体问题　three-body problem
三维的　three-dimensional
三维矢量　trivector
三相点　triple point
三相交变电流　three-phase alternating current
三相临界点　tricritical point
三相曲线　triple curve
三斜的　triclinic
三斜晶体　triclinic crystal
三斜晶系　triclinic system
三斜晶系的　triclinic
三原色视觉　trichromatic vision
散光　astigmatism
散焦　defocusing
散焦光束　defocused beam
散焦射束　defocused beam
散焦像　defocused image
散粒噪声　Schottky noise
散射　scattering
散射本领　scattering power
散射长度　scattering length
散射常数　scattering constant
散射光　scattered light
散射函数　scattering function
散射核　scattering kernel; scattering nucleus
散射恒量　scattering constant
散射角　scattering angle
散射截面　scattering cross-section
散射阱　scatter trap
散射矩阵　S-matrix (scattering matrix)
散射粒子　scattering particles
散射器　scatterer
散射损失　scattering loss
散射体　scatterer
散射图样　scattering pattern
散射仪　scatterometer
散度　div (divergence)
散裂　spallation
散落物　fallout
散热剂　coolant
扫描　scanning
扫描场平度　flatness of field

扫描成像 scanned imagery
扫描单色仪 scanning monochromator
扫描电子显微镜 SEM (scanning electron microscope)
扫描干涉仪 scanning interferometer
扫描激光器 scanned laser
扫描角 scanning angle
扫描孔径 scanning aperture
扫描棱镜 scanning prism
扫描器 scanner
扫描束 scanning beam
扫描隧穿显微镜 STM (scanning tunnelling microscope)
扫描透镜 scanning lens
扫描线圈 sweep coil
扫描仪 scanner
扫描照相机 scanning camera
色标 colour code
色彩 hue
色层分离谱 chromatogram
色层分析法 chromatography
色差 chromatism
色差补偿目镜 compensating eyepiece
色调 hue
色度 chromaticity; chrominance
色度计 chromatometer; chromometer; colorimeter
色度图 chromaticity diagram
色度学 colorimetry
色光 coloured light
色荷 colour charge
色觉计 chromatometer
色夸克 coloured quark
色粒子 coloured particle
色量计 colorimeter
色码 colour code
色盘 colour disk
色偏振 chromatic polarization
色品 chromaticity
色品图 chromaticity diagram
色谱法 chromatography
色谱图 chromatogram
色谱仪 chromatograph
色球(层) chromosphere
色散 dispersion
色散本领 dispersive power
色散方程 dispersion equation
色散关系 dispersion relation
色散介质 dispersive medium
色散棱镜 dispersing prism
色散率 dispersive power
色散媒质 dispersive medium
色散曲线 dispersion curve
色视觉 colour vision
色温 colour temperature
色像差 chromatic aberration
色心激光器 colour centre laser
铯(原子)钟 caesium clock
瑟尔热导仪 Searle conduction apparatus
刹车踏板 brake pedal
栅极 grid
闪点 flash point
闪电 lightning
闪光灯 photoflash lamp
闪光光解 flash photolysis
闪光体 twinkler
闪烁 scintillation
闪烁测井计数器 scintilogger
闪烁计数器 scintillation counter; scintillometer
闪烁晶体 scintillation crystal
闪烁镜 scintilloscope
闪烁谱仪 scintillation spectrometer
闪烁器 scintillator
闪烁扫描 scintiscanning
闪烁扫描器 scintiscanner
闪烁体 scintillator
闪烁照相机 scintillation camera
闪烁照相术 scintiphotography
闪烁照相图 scintiphotogram
闪耀光栅 blazed grating
嬗变 transmutation
熵 entropy
熵产生 entropy production
熵判据 entropy criterion
熵平衡方程 entropy balance equation

熵增加 entropy increase
熵增加原理 principle of entropy increase
上界 upper bound
上夸克 up quark
烧结 sintering
少数载流子 minority carrier
少体问题 few-body problem
少子 minority carrier
少子浓度 minority carrier density
射程 range
射弹 projectile
射电天文学 radio astronomy
射电通信 radio communication
射电望远镜 radio telescope
射电星 radio star
射电星系 radio galaxy
射电源 radio source
射极输出器 emitter follower
射解作用 radiolysis
射流 fluid；jet（stream）
射流技术 fluidics
射频 RF（radio-frequency）
射频(波)谱 radio-frequency spectrum
射频加热 radio-frequency heating
射频振荡器 radio-frequency oscillator
射气 emanation
射束点 beam spot
射束孔径角 beam angle
射线 ray
射线化学 actinology
射线图 ray diagram
射线照相术 radiography
射影变换 projective transformation
摄动 perturbation
摄谱学 spectrography
摄谱仪 spectrograph
摄氏度 degree Celsius
摄氏温标 Celsius（temperature）scale；Celsius thermometric scale；centigrade scale
摄氏温度计 Celsius thermometer
摄像机 camera
摄影机 camera
摄影学 photography
伸长计 extensometer
伸长弹性 tensile elasticity
伸长性 tensility
伸力 tensility
砷化镓 GaAs（gallium arsenide）
砷化镓激光器 GaAs laser
深度非弹性碰撞 deep inelastic collision
深度非弹性散射 deep inelastic scattering
深能级 deep level
深能级暂态谱学 deep level transient spectroscopy
甚低频 VLF（very low frequency）
甚高频 VHF（very high frequency）
渗出 effusion
渗漏系数 leakage factor
渗碳 cementation
渗碳体 cementite
渗透 osmosis
渗透率 permeability
渗透性 permeability
渗透压 osmotic pressure
渗透作用 osmosis
渗锌粉 sherardizing
渗压计 osmometer
蜃景 mirage
升 L；l（litre）
升华 sublimation
升华热 heat of sublimation
升华物质 sublimate
升力 ascensional force；lift force
升频转换 frequency upconversion
升压变压器 step-up transformer
生石灰 calx
生铁 cast iron；pig iron
生物发光 bioluminescence
生物工艺学 biotechnology

生物力学 biomechanics
生物能学 bioenergetics
生物物理学 biophysics
生物显微镜 biological microscope
生物荧光 bioluminescence
声 sound
声波 acoustical wave；sound wave
声波的 acoustic
声波雷达 sound radar
声导 acoustical conductance
声导率 acoustical conductivity
声的 acoustic
声电效应 acoustoelectric effect
声电装置 acoustoelectronic device
声调 intonation
声发射 acoustic emission
声光调制 acoustooptic modulation
声光效应 acoustooptic effect
声光学 acoustooptics
声级计 sound level meter
声阱 noise trap
声觉 acoustical perception
声抗 acoustical reactance
声脉冲 sound pulse
声呐 sonar (sound navigation and ranging)
声呐发射器 underwater sound projector
声纳 acoustic susceptance
声能学 sonics
声频范围 audiofrequency region
声频声子 acoustic phonon
声频振荡器 audio oscillator
声谱 sound spectrum
声谱计 acoustical spectrometer
声谱学 acoustical spectroscopy
声强级 sound-intensity level
声强计 acoustimeter；phonometer
声全息(照相) acoustical holography
声速 acoustic velocity；sonic speed；speed of sound
声透镜 acoustical lens
声线 sound ray
声学 acoustics
声学低温计 acoustic thermometer
声压 sound pressure
声掩蔽听力图 noise audiogram
声音 sound
声源 sound source
声障 sonic barrier
声震 sonic boom
声致发光 sonoluminescence
声子 phonon
声子谱 phonon spectrum
声阻 acoustical resistance
声阻抗 acoustic impedance
剩磁 remanence；retentivity
剩余极化强度 remanent polarization
剩余像差 residual aberration
失稳分解 spinodal decomposition
失相 out of phase
失真 deformation；distortion
失重 weightlessness
施轰粒子 bombarding particle
施密特校正板 Schmidt corrector (plate)
施密特照相机 Schmidt camera
施特恩-格拉赫实验 Stern-Gerlach experiment
施瓦氏半径 Schwarzschild radius
施瓦氏解 Schwarzschild solution
施温格效应 Schwinger effect
施主 donor
施主电离能 donor ionization energy
施主浓度 donor density
施主-受主对发光 donor-acceptor pair luminescence
湿度 humidity
湿度计 hygrometer
湿度调节器 humidistat
湿润器 humidifier

十进电阻箱 decade resistance box
十进计数管 dekatron
十进位制 decimal system
十亿 billion
十亿电子伏特 BeV（billion-electronvolt）；GeV（gigaelectronvolt）
十字线 graticule
石墨减速剂 graphite moderator
石屑 chip
石英 quartz
石英振荡器 quartz oscillator
石英钟 quartz clock
时标 time scale
时标因子 time scale factor
时差 equation of time
时基 time base
时间 time
时间反演 time reversal
时间反转 time reversal
时间扩展 time dila(ta)tion
时间膨胀 time dila(ta)tion
时间扫描 time sweep
时间矢量 time vector
时间相干性 temporal coherence
时间响应 time response
时间序列 time series
时间延缓 time dila(ta)tion
时距关系线图 distance-time graph
时距图 distance-time graph
时空 spacetime
时空均匀性 uniformity of spacetime
时空连续统 spacetime continuum
时空曲率 curvature of spacetime
时区 time zone
时效硬化 age-hardening
时序 sequence
实际输出功率 actual output
实际值 actual value
实物幻灯机 epidiascope
实像 real image
实验 experiment
实验 Q 值 experimental Q-value
实验室尺度 laboratory scale
实验物理学 experimental physics
实座 filled site
食 eclipse
矢积 vector product
矢量 vector（quantity）
矢量场 vector field
矢量分析 vector analysis
矢量积 vector product
矢量计 vector meter
矢量介子 vector meson
矢量流 vector current
矢量密度 vector density
矢量三角形 triangle of vectors
矢量图 vector diagram
矢通量 vector flux
矢性介子 vector meson
示波器 oscillograph；oscilloscope
示零器 null indicator
示温熔锥 pyrometric cone
示踪 labelling
示踪技术 tracer technique
示踪同位素 tracer isotope
世界线 world line
势 potential
势差 p.d.（potential difference）
势场 potential field
势函数 potential function
势降 potential drop
势阱 potential trough
势垒 （potential）barrier；potential hill
势垒穿透 barrier penetration
势垒高度 barrier height
势垒效应 barrier effect
势能 potential energy
势能面 potential energy surface
势散射 potential scattering
事件间隔 interval of events
事件穹界 event horizon
事件视界 event horizon
试探粒子 test particle

视差 parallax
视差角 parallactic angle
视差镜术 parallactoscopy
视差椭圆 parallactic ellipse
视差位移 parallactic displacement
视差误差 parallactic error
视差修正 parallax correction
视差移动 parallactic shift
视场 field of view；viewing field
视场光阑 field stop；view stop
视场深度 depth of field
视见度 visibility
视见函数 vision function
视觉 vision
视觉函数 vision function
视觉亮度 visual brightness
视觉暂留 persistence of vision
视力 vision
视膨胀 apparent expansion
视频显示装置 VDU（visual-display unit）
视双星 optical double star
视网膜 retina
视线 line of sight
视像管 vidicon
视星等 apparent stellar magnitude
视直径 visual diameter
适暗视觉 scotopic vision
适亮视觉 photopic vision
适配器 adapter
收气剂 getter
收音机 radio（receiver）
手柄 lever
手套箱 glove box
手性 chirality
手征荷 chiral charge
手征化合物 chiral compound
手征流 chiral current
手征性 chirality
守恒律 conservation law
守恒系 conservative system
首数 characteristic
寿命 lifetime
受导波 guided wave
受激二聚物 excimer
受激发射 stimulated emission
受激辐射 stimulated radiation
受激拉曼散射 stimulated Raman scattering
受激态 excited state
受激吸收 stimulated absorption
受激(准分)子 excimer
受迫发射 induced emission
受迫光散射 forced light scattering
受迫振动 forced vibration
受驱振荡 driven oscillation
受驱阻尼振子 driven damped oscillator
受主 acceptor
受主电离能 acceptor ionization energy
受主能级 acceptor（energy）level
受主浓度 acceptor density
受主杂质 acceptor impurity
枢轴 pivot
舒勒摆 Schuler pendulum
疏散星团 galactic cluster；open cluster
疏水 drain
输出电阻 output resistance
输出(额) output
输出功率 output power
输出功率计 output meter
输出量 output quantity
输出脉冲 output pulse
输出能量 output energy
输出耦合 output coupling
输出阻抗 output impedance
输电 electric power transmission
输电线 power line
输入设备 input device
输运方程 transport equation
输运理论 transport theory
输运系数 transport coefficient
输运现象 transport phenomenon
熟铁 wrought iron
熟铜 Muntz metal

鼠标 mouse
束 beam
束斑 beam spot
束半径 beam radius
束比 beam ratio
束缚电荷 bound charge
束缚电子 bound electron
束缚态 bound state
束孔 beam hole
束流强度 beam intensity
束强比 beam ratio
束腰 beam waist
束腰半径 waist radius
数据 data
数据采集 data acquisition
数据处理 data processing
数据库 database
数据容量 data capacity
数据舍弃 rejection of data
数理物理学 mathematical physics
数量积 scalar product
数量级 order of magnitude
数位 digit
数学摆 mathematical pendulum
数值孔径 numerical aperture
数字 digit
数字读出辉光管 digitron
数字伏特计 digital voltmeter
数字计时器 digital timer
数字计算机 digital computer
数字录音带 DAT（digital audio tape）
数字录制 digital recording
数字频率计 digital frequency meter
数字全息图 digital hologram
数字式多用表 digital multimeter
数字显示 digital display
数字指示管 digitron
刷形放电 brush discharge
衰变 decay
衰变产物 decay product
衰变常量 decay constant
衰变定律 decay law
衰变方式 decay modes
衰变恒量 decay constant
衰变宽度 decay width
衰变律 decay law
衰变率 decay rate
衰变能 decay energy
衰变图 decay modes
衰耗 attenuation
衰减 attenuation；damping；decay
衰减波 decaying wave
衰减率 decay rate
衰减系数 attenuation coefficient；damping coefficient
衰减振动 damped vibration
衰落 decay
双β衰变 double β-decay
双μ子 dimuon
双凹透镜 biconcave lens；double concave lens
双臂谱仪 double-arm spectrometer
双边约束 bilateral constraint
双侧约束 bilateral constraint
双掺杂激光器 double-doped laser
双重态 doublet
双重线 doublet
双重星系 double galaxy
双重子 dibaryon
双二极管 twin diode
双分子复合 bimolecular recombination
双峰势垒 double-humped barrier
双缝干涉 double-slit interference
双缝衍射 double-slit diffraction
双共振 double-resonance
双光栅摄谱仪 dual-grating spectrograph
双光束干涉 double-beam interference
双光束干涉测量术 two-beam interferometry
双光束干涉量度学 two-beam interferometry

双光子吸收 two-photon absorption
双光子相干态 two-photon coherent state
双合透镜 doublet
双荷子 dyon
双极场效应晶体管 nesistor
双极化子 bipolaron
双极性的 ambipolar
双极性等离子体 ambiplasma
双胶透镜 cemented doublet
双焦透镜 bifocal lens
双金属片 bimetallic strip
双金属温度计 bimetallic thermometer
双晶体 bicrystal; twin crystal
双聚焦谱仪 double-focusing spectrometer
双夸克 diquark
双棱镜 biprism
双脉冲编码 twin pulse code
双目镜 binoculars
双目视觉 binocular vision
双目望远镜 binocular field glasses; binocular telescope
双目显微镜 binocular microscope
双腔激光器 dual-cavity laser
双轻子 bilepton
双曲函数 hyperbolic function
双曲线 hyperbola
双栅场效(应)管 double-gate FET (double-gate field effect transistor)
双生子佯谬 twin paradox
双束分光光度计 double-beam spectrophotometer
双筒镜 binoculars
双筒望远镜 binocular field glasses; binocular telescope
双筒显微镜 binocular microscope
双凸透镜 biconvex lens; double convex lens
双椭圆腔 double-elliptical cavity
双微分截面 double-differential cross-section
双稳态电路 bistable circuit
双稳态多谐振荡器 flip-flop
双稳性 bistability
双线摆 bifilar pendulum
双线性耦合 bilinear coupling
双像棱镜 double-image prism
双芯簇射 double-core shower
双星 binary stars
双性杂质 amphoteric impurity
双压电晶片元件 bimorph cell
双原子气体 diatomic gas
双折射 birefringence; double refraction
双轴晶体 biaxial crystal
霜 frost; hoarfrost
水泵 water pump
水表 water meter
水波 water wave
水电 hydroelectricity
水合发光 lyoluminescence
水冷却 water cooling
水力 water power
水力电气 hydroelectricity
水力发动机 water engine; water motor
水力学 hydraulics
水能 water power
水平面 water level
水平投影 horizontal projection
水平轴 transverse axis
水汽密度 vapour density
水汽凝结体 hydrometeor
水汽现象 hydrometeor
水深计 fathometer
水位 (water) level
水位表 water level gauge
水位线 water line
水下摄影术 underwater photography
水下声标 underwater sound projector
水下视觉 underwater vision
水下探测 underwater probing
水星 Mercury
水压 hydraulic pressure

水压机 hydraulic press；water engine
水银 mercury
水银灯 mercury lamp
水银气压计 mercury barometer
水银柱 mercury column
水蒸气 water vapour
水准器 level
水准仪 water level（gauge）
顺磁波子 paramagnon
顺磁磁化率 paramagnetic susceptibility
顺磁共振 PMR（paramagnetic resonance）
顺磁矩 paramagnetic moment
顺磁性 paramagnetism
顺磁振子 paramagnon
顺行 direct motion
瞬变电流 transient current
瞬变平衡 transient equilibrium
瞬变响应 transient response
瞬变运动 transient motion
瞬发中子 prompt neutron
瞬时速度 instantaneous velocity
瞬时值 instantaneous value
瞬时中心 instantaneous centre
瞬时转动轴 instantaneous axis of rotation
瞬时状态 transient state
瞬态 transience；transient state
瞬态脉冲 transient pulse
瞬态响应 transient response
瞬态运动 transient motion
瞬态振动 transient vibration
瞬心 instantaneous centre
瞬子 instanton
朔望 syzygy
朔望潮 spring tide
朔望月 lunation；synodic month
丝极 filament
斯(勒格) slug
斯涅耳定律 Snell's law
斯涅耳折射定律 Snell's law of refraction
斯塔克效应 Stark effect
斯特藩-玻尔兹曼定律 Stefan-Boltzmann law
斯特藩定律 Stefan's law
斯特林热气机 Stirling engine
斯特林循环 Stirling cycle
斯托克斯参数 Stokes' parameters
斯托克斯定理 Stokes' theorem
斯托克斯定律 Stokes' law
死时间 dead time
四倍频器 quadrupler
四冲程循环 four-stroke cycle
四冲程引擎 four-stroke-cycle engine
四方晶系 tetragonal system
四极管 quadrode；tetrode
四极矩 quadrupole moment
四极子 quadrupole
四角晶系 tetragonal system
四面体 tetrahedron
四能级系统 four-level system
四维动量 four-momentum
四维加速度 four-acceleration
四维力 four-force
四维时空 four-dimensional spacetime
四维矢量 four-vector
四维速度 four-velocity
四维形式〖麦克斯韦方程的〗 four-dimensional form of Maxwell equations
四维张量 four-tensor
四元矢量 four-vector
四元速度 four-velocity
四轴坐标系 four-axis coordinate system
搜索 hunting
速度 speed；velocity
速(度)比 velocity ratio
速度表 velocimeter
速度场 velocity field
速度的平行四边形 parallelogram of velocities
速度分解 resolution of velocity
速度合成 composition of velocities
速度计 velocimeter

速度空间 velocity space
速度约束 constraint of velocity
速率 speed
速率计 speedometer
速率-时间图 speed-time graph
速调管 klystron
塑性 plasticity
塑性晶体 plastic crystal
塑性力学 plasticity
塑性形变 plastic deformation
酸 acid
酸性转炉炼钢法 Bessemer process
算法 algorithm
算法复杂性 algorithm complexity
算符 operator
算术级数 arithmetic progression; arithmetic series
算术逻辑单元 ALU (arithmetic logic unit)
算术平均 arithmetic mean
算术平均数 arithmetic average; arithmetic mean
算术平均值 arithmetic average; arithmetic mean
算子 operator
随机存取存储器 RAM (random-access memory)
随机过程 stochastic process
随机误差 stochastic error
随机游动 random walk
随机运动 random motion
随遇平衡 indifferent equilibrium; neutral equilibrium
碎裂反应 fragmentation reaction
隧道电流 tunnel current
隧道二极管 Esaki diode; tunnel diode
隧道光电二极管 tunnel photodiode
隧道结 tunnel junction
隧道晶体管 tunnel transistor
隧道效应 tunnel effect
缩放仪 pantograph
缩微胶片 microfilm
缩微像存储 microimage storage
缩微照片 microphotograph
索末菲椭圆轨道 Sommerfeld elliptic orbit
锁模 mode-locking
锁频 frequency-locking
锁相 phase-locking

Tt

台阶　plateau
太空飞行器　spacecraft
太阳　sun
太阳常数　solar constant
太阳潮　solar tide
太阳磁(学)　solar magnetism
太阳电池板　solar panel
太阳风　solar wind
太阳辐射　solar radiation
太阳光谱　solar spectrum
太阳黑子　sunspot
太阳恒量　solar constant
太阳能　solar energy
太阳能电池　solar cell
太阳年　solar year
太阳日　solar day
太阳入射辐射　incoming solar radiation
太阳视差　solar parallax
太阳望远镜　solar telescope
太阳物理学　solar physics
太阳系　solar system
太阳耀斑　solar flare
太阴月　lunation
态　state
态变量　state variable
态参量　state parameter
态函数　state function
态空间　state space
态矢量　state vector
泰勒级数　Taylor series
坍缩　collapse
坍(缩恢)复现象　collapse-revivals
坍缩星　collapsar
弹光效应　elasto-optic effect
弹簧　spring
弹簧秤　spring balance
弹力　elastic force; elasticity
弹性　elasticity
弹性变形　elastic deformation
弹性力　elastic force
弹性模量　elastic modulus; modulus of elasticity
弹性模数　elastic modulus; modulus of elasticity
弹性能　elastic energy
弹性碰撞　elastic collision
弹性散射　elastic scattering
弹性散射截面　elastic scattering cross-section
弹性体　elastic body
弹性学　elasticity
弹性应力　elastic stress
探测　detection; tracking
探测灵敏度　detection sensitivity
探测器　detector
探测器分辨率　detector resolution
探测效率　detection efficiency
探测仪　finder
探察线圈　search coil
探空仪　sonde
探头　sonde
探针　sonde
探针技术　probe technique
碳化铁　cementite
碳年代法　carbon dating
碳丝　carbon fibres
碳纤维　carbon fibres
碳循环　carbon cycle
汤川相互作用　Yukawa interaction
汤姆逊散射　Thomson scattering

汤姆逊效应 Thomson effect
逃逸 escapement
逃逸深度 escape depth
逃逸速度 escape velocity
淘析 elutriation
淘洗 elutriation
特超声学 hypersonics
特定相 specific phase
特高频 UHF（ultrahigh frequency）
特殊恒星 peculiar star
特斯拉 T（tesla）
特斯拉变压器 Tesla transformer
特斯拉线圈 Tesla coil
特性函数 characteristic function
特性曲线 characteristic curve
特征 X 射线 characteristic X-ray
特征方程 characteristic equation；secular equation
特征光谱 characteristic spectrum
特征函数 eigenfunction；proper function
特征频率 characteristic frequency
特征数 characteristic
特征向量 eigenvector
特征值 eigenvalue；proper value
特征状态 characteristic state
梯度 gradient
梯度算子 gradient operator
梯度展开 gradient expansion
梯图 ladder diagram
梯图近似 ladder approximation
锑化铟 indium antimonide
提纯 purification
提取 extraction
体电荷 volume charge
体积 volume
体积分 volume integral
体积计 stereometer；volumenometer；volumeter
体积力 volume force
体积模量 bulk modulus
体积膨胀 cubical expansion；volume expansion
体积速度 volume velocity
体视镜 stereoscope
体视望远镜 relief telescope；stereotelescope；telestereoscope
体温计 clinical thermometer
体心点阵 body-centred lattice
体心格 body-centred lattice
体心立方体 b. c. c.（body-centred cubic）
替代能源 alternative energy
天赤道 celestial equator
天底 nadir
天顶 zenith
天空电波 sky wave
天平 balance
天平动 libration
天球 （celestial）sphere
天然磁石 lodestone
天然放射性 natural radioactivity
天体 celestial body
天体测量学 astrometry
天体光谱学 astrospectroscopy
天体力学 celestial mechanics
天体望远镜 celescope
天体物理学 astrophysics
天体照相术 celestial photography
天体照相仪 photographic telescope
天图 star atlas
天文单位 AU（astronomical unit）
天文光学 astronomical optics
天文台 observatory
天文望远镜 astronomical telescope
天文学 astronomy
天线 aerial
天线杆 aerial mast
天线耦合 aerial coupling
天线调谐 aerial tuning
天线效应 aerial effect
天线增益 aerial gain
天线阵 aerial array
填充密度 packing density

填隙式化合物 interstitial compound
填隙原子 interstitial
填隙子 interstitial
条纹 fringe
条纹衬比度 fringe contrast
条纹反差 fringe contrast
条纹可见度 fringe visibility
调 Q Q-modulation
调板电路 tuned plate circuit
调幅 AM（amplitude modulation）
调幅光 amplitude modulated light
调和级数 harmonic progression；harmonic series
调焦 accommodation；focus
调节 accommodation；adjustment
调频 FM（frequency modulation）
调栅电路 tuned grid circuit
调速 velocity modulation
调速器 governor
调相 phase modulation
调谐 tuning
调谐电路 tuned circuit
调谐放大器 tuned amplifier
调谐器 tuner
调谐特性 tuning characteristic
调谐线圈 tuning coil
调整 adjustment
调直 straightening
调制 modulation
调制传递函数 MTF（modulation transfer function）
调制光谱学 modulation spectroscopy
调制解调器 modem（modulator-demodulator）
跳变 jump
铁磁波子 ferromagnon
铁磁玻璃 ferromagnetic glass
铁磁材料 ferromagnetic material
铁磁共振 FMR（ferromagnetic resonance）
铁磁居里点 ferromagnetic Curie point
铁磁体 ferromagnet
铁磁相变 ferromagnetic phase transition
铁磁性 ferromagnetism
铁磁振子 ferromagnon
铁电材料 ferroelectric material
铁电畴 ferroelectric domain
铁电畴壁 ferroelectric domain wall
铁电传感器 ferroelectric sensor
铁电的 ferroelectric
铁电体 ferroelectric
铁电铁磁体 ferroelectric ferromagnet
铁电现象 ferroelectricity
铁电性 ferroelectricity
铁合金 ferroalloy
铁素体 ferrite
铁弹畴 ferroelastic domain
铁弹体 ferroelastics
铁弹性 ferroelasticity
铁芯 slug
铁氧体 ferrite
铁氧体磁性 ferrimagnetism
听度计 audiometer
听觉 acoustical perception
听力计 audiometer
廷德尔效应 Tyndall effect
停止 stop
通带 passband；transmission band
通道 channel
通孔 through hole
通量 flux
通量密度 flux density
通路容量 channel capacity
通信卫星 communication satellite；telecommunication satellite
通用式电动机 universal motor
同步的 synchronous
同步电动机 synchronous motor
同步辐射 synchrotron radiation

同步轨道 geostationary orbit; geosynchronous orbit; stationary orbit; synchronous orbit
同步回旋加速器 synchrocyclotron
同步加速器 synchrotron
同步加速器辐射 synchrotron radiation
同步卫星 stationary satellite; synchronous satellite
同步稳相加速器 synchrophasotron
同步旋转 synchronous rotation
同分异构体 isomer
同分异构性 isomerism
同核异能素 isomer
同极电荷 homocharge
同晶型(性) isomorphism
同科电子 equivalent electron
同量异位素 isobar
同素异形体 allotrope
同素异形(现象) allotropism; allotropy
同位素 isotope
同位素分离 isotope separation
同位素分离器 isotron
同位素分析器 isotron
同位素丰度 isotopic abundance
同位素效应 isotope effect
同位素学 isotopy
同位素移位 isotope shift
同位旋 isobaric spin; isospin; isotopic spin
同温层 stratosphere
同向自旋配对态 ESP state (equal-spin-pairing state)
同相 in-phase
同消色线 isogyre
同心光束 concentric beam; homocentric beam
同心透镜 concentric lens
同形性 isomorphism
同一性 identity
同质结 homojunction
同质结激光器 homojunction laser
同质外延 homoepitaxy
同质异能素 isomer
同质异能性 isomerism
同质异位核 isobaric nucleus
同中子异位素 isotone
同轴电缆 coaxial cable
同轴线 coaxial cable
铜镍锌合金 German silver; nickel silver
铜锡合金 gun metal
瞳函数 pupil function
瞳孔 pupil
统计法 statistics
统计规律性 statistical regularity
统计力学 statistical mechanics
统计平衡 statistical equilibrium
统计平均(值) statistical average
统计权重 statistical weight
统计物理学 statistical physics
统计误差 statistical error
统计学 statistics
统一场论 unified-field theory
统一资源定位符 URL (universal resource locator)
桶形畸变 barrel distortion
桶形失真 barrel distortion
痛阈 threshold of feeling; threshold of pain
头波 bow wave
投光灯 projector
投影 projection
投影算符 projection operator
投影物镜 projection objective
投影仪 projector
投掷 projection
透度计 penetrometer
透镜 lens
透镜分光计 lens spectrometer
透镜公式 lens formula
透镜光心 lens centre
透镜谱仪 lens spectrometer
透镜中心 lens centre
透镜组 combination of lenses; lens combination
透明度 transparency
透明性 transparency

透热性 diathermancy
透射 transmission
透射比 transmittance
透射电子显微镜 transmission electron microscope
透射光栅 transmission grating
透射计 transmissometer
透射率 transmissivity
透射膜 transmission film
透射实验 transmission experiment
透射系数 transmission coefficient；transmittance
透射照相机 transmission camera
凸面的 convex
凸面镜 convex mirror
凸透镜 convex lens
凸月 gibbous Moon
突变 catastrophe
突变结 abrupt junction
突变理论 catastrophe theory
突变异质结 abrupt heterojunction
突跳 jump
图 graph
图解法 graphical method
图解展开 graphical expansion
图灵机 Turing machine
图像 image
图像编码 image encoding
图像变换 image conversion
图像变形 anamorphose
图像处理 image processing
图像存储 image storage
图像分解力 image definition
图像复原 image restoration
图像合成 image synthesis
图像模糊 image blurring
图像清晰度 image definition
图像去模糊 image deblurring
图像失真 anamorphose
图像增强 image enhancement
图像综合 image synthesis
涂膜 blooming
钍 Th（thorium）
钍系 thorium series
湍动 turbulent motion
湍动激波 turbulent shock
湍动加热 turbulent heating
湍动谱 turbulent spectrum
湍流 turbulence；turbulent flow
湍流分离 turbulent separation
湍流混合 turbulent mixing
湍流激波 turbulent shock
湍流加热 turbulent heating
湍流脉动 turbulent fluctuation
湍流谱 turbulent spectrum
湍流起伏现象 turbulent fluctuation
湍流强度 turbulence intensity
湍流区域 turbulent region
湍流通量 turbulent flux
湍流运动 turbulent motion
湍流涨落现象 turbulent fluctuation
湍流阻力 turbulent resistance
湍射流 turbulent jet
湍性媒质 turbulent medium
推迟板 retardation plate
推斥力 repulsive force
推进器 propeller
推力 thrust
推挽式的 push-pull
推挽式放大器 push-pull amplifier
退磁 degaussing；demagnetization
退磁因数 demagnetization factor
退化度 degeneracy
退化态 degenerate states
退化系 degenerate system
退火 annealing；tempering
退激发 deexcitation
退极化 depolarization
退极化因子 depolarization factor
退禁闭 deconfinement
退耦参量 decoupling parameter
退耦带 decoupled band
退偏振 depolarization
蜕变 disintegration；

transmutation
蜕变常数 disintegration constant
蜕变恒量 disintegration constant
托 torr
托卡马克 Tokamak
托勒密体系 Ptolemaic system
托里拆利真空 Torricellian vacuum
脱附 desorption
脱溶 precipitation
脱色 decoloration
脱逸电子 runaway electron
陀螺仪 gyroscope
椭球面 ellipsoid
椭球面镜 ellipsoidal mirror
椭球腔 ellipsoidal cavity
椭圆 ellipse
椭圆偏振 elliptic polarization
椭(圆)偏(振)测量术 ellipsometry
椭(圆)偏(振)计 ellipsometer
椭圆星系 elliptical galaxy
拓扑学 topology

Ww

瓦　W (watt)
瓦时计　watthour meter
瓦特　W (watt)
瓦特计　wattmeter
瓦特时　watthour
瓦特数　wattage
外差(法)的　heterodyne
外大气层　exosphere
外反射　external reflection
外赋半导体　extrinsic semiconductor
外赋吸收　extrinsic absorption
外力　external force
外设　peripheral device
外推　extrapolation
外围设备　peripheral device
外消旋混合物　racemate; racemic mixture
外延　epitaxy
外延沉积　epitaxial deposition
外延生长　epitaxial growth
外逸层　exosphere
弯电效应　flexoelectric effect
弯(曲)矩　bending moment
弯曲时空　curved spacetime
弯曲应变　bending strain
弯曲应力　bending stress
弯月面　meniscus
弯月形透镜　meniscus
完备集　perfect set
完美晶体　perfect crystal
完美溶液　ideal solution
完全非弹性碰撞　completely inelastic collision; perfect inelastic collision
完全辐射体　perfect radiator
完全滚动　pure rolling
完全黑体　perfect black body
完全抗磁性　perfect diamagnetism
完全气体　perfect gas
完全弹性碰撞　perfect elastic collision
完全吸收体　perfect absorber
完全相干光　completely coherent light
完全相干性　full coherence
顽磁　remanence
顽磁性　retentivity
万维网　Web; WWW (World Wide Web)
万向节　gimbal
万用表　multimeter; universal meter
万用电表　avometer
万有引力　gravitational force; universal gravitation
万有引力定律　law of universal gravitation
网格　raster
网格法　cellular method
望远镜　telescope
望远镜(制造)学　telescopy
望远瞄准器　telescopic sight
望远显微镜　telemicroscope
威尔逊云室　Wilson cloud chamber
微安　μA (microampere)
微巴　barye
微变电容器　vernier condenser
微变聚焦　vernier focusing
微波　microwave
微波背景辐射　microwave background radiation

微波测距仪 tellurometer
微波光学 microwave optics
微波激射(器) maser
微波三极管 dyotron
微处理器 microprocessor
微电子学 microelectronics
微动 perturbation
微分法 differentiation
微分方程 differential equation
微分几何 differential geometry
微分散射截面 differential scattering cross-section
微分学 differential calculus
微伏 microvolt
微观的 microscopic
微观粒子 microscopic particle
微观量 microscopic quantity
微观世界 microworld
微观态 microscopic state
微观物理学 microphysics
微观宇宙 microcosm
微光视觉 scotopic vision
微积分 calculus
微结构 microstructure
微晶(体) crystallite; microcrystal
微居里 μCi (microcurie)
微粒论 particle formalism
微粒说 corpuscular theory
微粒性 corpuscular property
微量天平 microbalance
微米 μm (micron)
微缺陷 microdefect
微扰 perturbation
微扰论 perturbation theory
微商 derivative
微调 fine adjustment
微调装置 vernier device
微小陨石 micrometeorite
微陨星 micrometeorite
韦伯 Wb (weber)
韦斯顿电池 Weston cell
唯一性定理 uniqueness theorem
维德曼-弗兰兹定律 Wiedemann-Franz law
维度 dimension
维恩公式 Wien formula
维恩图解 Venn diagram
维恩位移(定)律 Wien's displacement law
维格纳核素 Wigner nuclides
维格纳能量 Wigner energy
维格纳-威特默定则 Wigner-Witmer rules
维里方程 virial equation
维姆胡斯特起电机 Wimshurst machine
伪矢量 pseudovector
伪像 spurious image
伪真空 false vacuum
尾数 mantissa
纬度 latitude
纬线 latitude
卫铁 keeper
卫星 satellite
未饱和汽 unsaturated vapour
未激活态 unactivated state
未来 future
位 potential
位差 p.d. (potential difference)
位场 potential field
位错 dislocation
位错滑移 slip of dislocation
位函数 potential function
位降 potential drop
位阱 potential trough
位垒 potential barrier; potential hill
位矢 position vector
位势论 potential theory
位相 phase
位相散乱 phase debunching
位形 configuration
位形空间 configuration space
位移 displacement
位移电流 displacement current
位置色差 chromatism of position
位置矢量 position vector
位置约束 constraint of position
味 flavour

味对称性 flavour symmetry
味空间 flavour space
温标 temperature scale; thermometric scale
温伯格-萨拉姆模型 WS model (Weinberg-Salam model)
温差 temperature difference
温差电 thermoelectricity
温差电堆 thermopile
温差电偶 thermocouple
温差电偶真空规 thermocouple gauge
温差电效应 thermoelectric effect
温度 temperature
温度变化 temperature variation
温度补偿 temperature compensation
温度补偿器 temperature compensator
温度测定法 thermometry
温度场 temperature field; thermal field
温度峰值 thermal spike
温度计 thermometer
温度记录器 thermograph
温度记录术 thermography
温度降差 temperature drop
温度校正 temperature correction
温度弥散 temperature dispersion
温度漂移 temperature drift
温度气压表 thermobarometer
温度梯度 temperature gradient
温度系数 temperature coefficient
温熵图 tephigram
温室效应 greenhouse effect
文件 file
文丘里管 Venturi tube
文氏管 Venturi tube
纹影摄影术 schlieren photography
紊流 turbulence; turbulent flow
稳定轨道 stationary orbit
稳定粒子 stable particle
稳定流动 streamline flow
稳定平衡 stable equilibrium; stationary equilibrium
稳定系综 stationary ensemble
稳定性 stability
稳定性判据 stability criterion
稳定性准则 stability criterion
稳恒电流 steady current
稳恒态 steady state
稳流 steady flow
稳流电源 stabilized current supply
稳流电阻器 ballast resistor
稳频 frequency stabilization
稳频激光器 frequency stabilized laser
稳态 stationary state
稳相 stationary phase
稳相加速器 phasotron
稳压电源 stabilized voltage supply
稳压二极管 voltage stabilizing diode; Zener diode
涡电流 eddy current
涡度 vorticity
涡对 vortex pair
涡流 bow wave; eddy current; turbulent flow; vortex flow
涡流带 vortex strip
涡流损耗 eddy current loss
涡轮(机) turbine
涡轮喷气发动机 turbojet
涡轮式发电机 turbogenerator
涡线对 vortex pair
涡旋 vortex
涡旋强度 vorticity
涡旋稳定 vortex stabilization
沃拉斯顿棱镜 Wollaston prism
污染 pollution
无差拍 deadbeat
无场空间 field-free space
无定向磁强计 astatic magnetometer
无定向电流计 astatic galvanometer
无定形的 amorphous

无定形体 amorphous solid
无辐射衰减 radiationless decay
无辐射跃迁 radiationless transition
无公度相 incommensurate phase
无功电流 reactive current
无功功率 reactive power
无关参量 irrelevant parameter
无关性 independence
无规行走 random walk
无规运动 random motion
无畸变目镜 orthoscopic eyepiece
无畸变图像 orthoscopic image; undistorted image
无极分子 nonpolar molecule
无极化合物 nonpolar compound
无焦成像系统 afocal imaging system
无理数 irrational number; surd
无力场 force-free field
无量纲化 nondimensionalization
无毛定理 no-hair theorem
无黏流动 nonviscous flow
无碰撞不稳定性 collisionless instability
无碰撞阻尼 collisionless damping
无穷大 infinity
无穷级数 infinite series
无穷小的 infinitesimal
无缺陷晶体 perfect crystal
无色水晶 rock crystal
无声放电处理 voltolization
无损检测 nondestructive inspection
无透镜傅里叶全息图 lensless Fourier hologram
无吸收材料 absorption-free material
无限 infinity
无限小的 infinitesimal
无线电 radio
无线电报(学) wireless telegraphy
无线电传递 radio transmission
无线电窗 radio window
无线电电子学 radioelectronics
无线电定位 radiolocation
无线电干涉仪 radio interferometer
无线电广播 radio broadcasting
无线电话学 wireless telephony
无线电接收机 radio receiver
无线电探空仪 radiosonde
无线电天文望远镜 radio telescope
无线电通信 radio communication
无线电源 radio source
无向量 scalar (quantity)
无像差系统 aberration-free system
无序 disorder
无旋流模型 irrotational flow model
无液气压计 aneroid barometer
无源反射器 parasitic reflector
无源器件 passive device
无源元件 parasitic element
无重力环境 gravity-free environment
无周期性 aperiodicity
五极管 pentode
伍德合金 Wood's metal
物 object
物端光栅 objective grating
物端棱镜 objective prism
物方焦点 focus in object space
物方空间 object space
物光束 object beam
物镜 objective (lens)
物镜调整 objective lens adjustment
物距 object distance
物空间 object space
物理摆 physical pendulum
物理光学 physical optics
物理化学 physical chemistry
物理量 physical quantity
物理天平 physical balance
物理吸附 physisorption

物理学 physics
物平面 object plane
物态 state of matter
物态变化 change of state
物态参量 state parameter
物态方程 equation of state
物体 object
物质 matter
物质波 matter wave
物质总量 amount of substance
误差 error
误差传递 propagation of error

Xx

西弗特　Sv (sievert)
西门子　S (siemens)
吸附　adsorption; sorption
吸附泵　sorption pump
吸附剂　adsorbent
吸附能力　adsorbability
吸附热　heat of adsorption
吸附原子　adatom
吸附座　adsorption site
吸积盘　accretion disk
吸留　occlusion
吸能的　endoergic
吸气剂　getter
吸热变暗　decalescence
吸热反应　endothermic reaction
吸声　sound absorption
吸声板　abat-voix
吸收　absorption
吸收本领　absorptive power
吸收比　absorptance; absorptivity
吸收度　absorptance
吸收峰　absorption peak
吸收光谱　absorption spectrum
吸收率　absorptivity
吸收能力　absorptive power
吸收频谱　absorption spectrum
吸收系数　absorption coefficient; absorptivity
吸引　attraction
吸引力　attraction; attractive force
吸引子　attractor
吸杂剂　getter
希尔伯特空间　Hilbert space
希尔德布兰德规则　Hildebrand rule
希格斯场　Higgs field
希格斯粒子　Higgs particle
硒光电池　selenium cell
稀薄等离(子)体　rarefied plasma
稀化　rarefaction
稀疏　rarefaction
稀疏波　rarefaction wave
稀土金属　rare-earth metal
稀土永磁体　rare-earth permanent magnet
稀有金属　rare metal
熄灭　extinction
洗出液　eluate
洗提　elution
洗提液　eluent
徙动速度　velocity of migration
系　train
系数　coefficient; factor
系统软件　systems software
系统误差　systematic error
系综　ensemble
系综理论　ensemble theory
细胞　cell
细粒密度　fine-grained density
细调　fine adjustment
细致平衡　detailed balancing
细致平衡原理　detailed balancing principle
隙能级　gap level
隙透　effusion
狭缝　slit
狭义相对论　restricted theory of relativity; special (theory of) relativity
狭义相对性原理　principle of special relativity
下夸克　d-quark

纤维 fibre
纤维丛 fibre bundle
纤维光导 fibre light guide
纤维光学 fibre optics
纤维镜 fibrescope
纤维束 fibre bundle
纤维质料 fibre
弦 string
弦潮 neap tide
弦模型 string model
弦音计 sonometer
衔铁 armature
舷波 bow wave
显微光度计 microphotometer
显微结构 microstructure
显微镜 microscope
显微镜的 microscopic
显微照片 microphotograph; photomicrograph
显微照相术 photomicrography
显像管 teletron
显像密度计 densitometer
现在 present
限度 limit
线电荷密度 linear charge density
线分辨率 linear resolution
线光谱 line spectrum
线光栅 wire grating
线积分 line integral
线接头 terminal
线宽 line breadth; line width
线路 circuit
线路图 circuit diagram
线密度 linear density
线能量转移 LET (linear energy transfer)
线膨胀率 linear expansivity
线偏振 linear polarization
线圈 coil
线缺陷 line defect
线栅 wire grating
线素 line element
线系 line series
线形 line shape
线性方程 linear equation
线性放大(率) linear magnification
线性放大器 linear amplifier
线性光学 linear optics
线性加速器 linac; linear accelerator
线性空间 linear space
线性能量转移 LET (linear energy transfer)
线性吸收系数 linear absorption coefficient
线性元件 linear element
线元 line element
线状谱 line spectrum
陷波电路 trap (circuit)
陷波器 trapper
陷俘 trap
陷俘粒子 trapped particle
陷俘粒子不稳定性 trapped particle instability
陷光器 light trap
陷获 trap
陷获效应 trapping effect
陷阱 trap
陷阱模型 trap model
陷模 trapped mode
相对磁导率 relative permeability
相对电容率 relative permittivity
相对加速度 relative acceleration
相对孔径 relative aperture
相对口径 relative aperture
相对论 relativity (theory)
相对论性场方程 relativistic field equation
相对论性的 relativistic
相对论(性)力学 relativistic mechanics
相对论性粒子 relativistic particle
相对论(性)量子力学 relativistic quantum mechanics
相对论(性)物理学 relativistic physics
相对论(性)效应 relativistic effect

相对论(性)质量 relativistic mass
相对密度 r.d.(relative density)
相对渗透率 relative permeability
相对湿度 relative humidity
相对速度 relative velocity
相对误差 relative error
相对性 relativity
相对性〖同时性的〗 relativity of simultaneity
相对性原理 principle of relativity; relativity principle
相对原子质量 r.a.m.(relative atomic mass)
相对运动 relative motion
相干波 coherent wave
相干长度 coherence length
相干成像 coherence imaging
相干度 degree of coherence
相干辐射 coherent radiation
相干光 coherent light
相干光源 coherent source
相干散射 coherent scattering
相干时间 coherence time
相干态 coherent state
相干条件 coherence condition
相干效应 coherence effect
相干性 coherence
相关函数 correlation function
相关系数 correlation coefficient
相互作用 interaction
相加色 additive colour
相消干涉 destructive interference
响度 loudness
响度级 volume level
响应函数 response function
响应曲线 response curve
向场镜 field lens
向量 vector (quantity)
向量的平行四边形 parallelogram of vectors
向量积 cross product; vector product
向量空间 vector space
向量流量 vector flux
向量密度 vector density
向列型晶体 nematic crystal
向心爆炸 implosion
向心加速度 centripetal acceleration
向心力 centripetal force
相 phase
相变 change of phase; phase change; phase transition
相变平衡条件 equilibrium condition of phase transition
相差 phase difference
相衬 phase contrast
相衬显微镜 phase-contrast microscope
相光栅 phase grating
相积分 phase integral
相畸变 phase distortion
相鉴频器 phase discriminator
相角 phase angle
相空间 phase space
相律 phase rule
相片 photograph
相平衡 phase equilibrium
相栅 phase grating
相矢量 phasor
相速(度) phase speed; phase velocity
相图 phase diagram
相位 phase
相位比较器 phase comparator
相位补偿 phase compensation
相位差 phase difference
相位常量 phase constant
相位超前 phase advance; phase lead
相位反差 phase contrast
相位分辨率 phase resolution
相位计 phase meter
相位鉴别器 phase discriminator
相位校正器 phase corrector
相位解调器 phase demodulator
相位匹配 phase matching
相位曲线 phase curve
相位失真 phase distortion
相位调制 phase modulation

相位跳变 phase jump
相位同步 phase synchronization
相位相干性 phase coherence
相位型光栅 phase grating
相位延迟 phase retardation
相位移 phase displacement
相位跃变 phase jump
相位滞后 phase lag
相移 phase displacement；phase shift
相坐标 phase coordinates
像 image
像差 aberration
像差校正 aberration correction
像差曲线 aberration curve
像场 image field
像场校正镜 field corrector
像场校正器 field corrector
像场平度 flatness of field
像场弯曲 curvature of field
像方焦点 focus in image space
像方空间 image space
像管 image intensifier
像距 image distance
像空间 image space
像平面 image plane
像散现象 astigmatism
像散性 astigmatism
像素 image element；pixel
像歪曲透镜 anamorphote lens
像元 image element；pixel
像增强器 image intensifier
像质 image quality
像主距 image distance
像转换器 image converter
肖克利能级 Shockley energy level
肖特基缺陷 Schottky defect
肖特基势垒 Schottky barrier
肖特基效应 Schottky effect
肖特基噪声 Schottky noise
消磁 degaussing
消光 extinction
消光比 extinction ratio
消光定理 extinction theorem
消光校正 extinction correction
消光率 extinction index
消光系数 extinction coefficient
消耗 dissipation
消偏振 depolarization
消偏振因数 depolarization factor
消球差透镜 aplanat；aplanatic lens
消色差 achromatic aberration
消色差三合透镜 achromatic triplet
消色差双合透镜 achromatic doublet
消色差透镜 achromat；achromatic lens
消色差性 achromaticity；achromatism
消声室 anechoic chamber
消像差透镜 anastigmat
消像差系统 aberration-free system
消像散器 stigmator
消像散透镜 anastigmat
小潮 neap tide
小风速计 airmeter
小角度碰撞 glancing collision
小角度入射 glancing incidence
小角散射 small-angle scattering
小阶梯光栅 echelette grating
小卡 therm
小面化 faceting
小面生长 facet growth
小时 hour
小行星 asteroid；minor planet；planetoid
小行星带 asteroid belt
小型双眼望远镜 opera glasses
效率 efficiency
啸叫 howl
楔形 wedge shape
协作现象 cooperative phenomenon
胁变 strain
胁变硬化 strain hardening
胁强 stress
胁强张量 stress tensor

斜点阵 oblique lattice
斜对称 skew symmetry
斜方晶系的 orthorhombic
斜光线 oblique ray
斜角 oblique angle
斜面 inclined plane；oblique plane；ramp
斜碰 oblique impact
斜坡 ramp
斜入射 oblique incidence
谐波 harmonic
谐波振荡器 harmonic oscillator
谐和 consonance
谐量图 vector diagram
谐音 harmonic (sound)
谐振 resonance
谐振腔 cavity resonator；resonant cavity
谐振子 harmonic oscillator
泄漏电阻 leakage resistance
泄气阀 air escape valve
泻流 effusion
心电图 ECG (electrocardiogram)
芯极化 core polarization
芯片 chip
辛烷数 octane number
辛烷值 octane number
锌 Zn (zinc)
锌粉热镀 sherardizing
新星 new star；nova
新星爆发 nova outburst
信道 (information) channel
信道容量 channel capacity
信号 signal
信息光学 information optics
信息容量 data capacity；information capacity
信息熵 information entropy
信噪比 signal-to-noise ratio
星 star
星等 (stellar) magnitude
星等差 magnitude difference
星际空间 interstellar space
星际气体 interstellar gas
星历表 ephemeris
星图 star atlas
星团 star cluster
星位角 parallactic angle
星系 galaxy
星系际空间 intergalactic space
星系体 galaxoid
星云 nebula；star cloud
星云假说 nebular hypothesis
行波 travelling wave
行星 planet
行星际空间 interplanetary space
行星际太空 interplanetary space
行星运动 planetary motion
行移场 travelling field
形变 deformation
形变参量 deformation parameter
形变参数 deformation parameter
形变能 deformation energy
形成热 heat of formation
形成时间 formation time
形成因子 formation factor
形心 centroid
形状记忆合金 shape-memory alloy
形状因子 form factor
性能因数 performance factor
修正 correction
虚功 virtual work
虚功原理 principle of virtual work
虚粒子 virtual particle
虚拟电荷 fictitious charge
虚设电荷 fictitious charge
虚数 imaginary number
虚态 virtual state
虚位移 virtual displacement
虚位移原理 principle of virtual displacement
虚像 virtual image
许可载荷 allowable load
许用偏差 allowable deviation
序参数 order parameter
序列 sequence
蓄电池 accumulator；secondary

cell；(storage) battery；storage cell
蓄电池组 storage battery
蓄能器 accumulator
悬浮 suspension
悬浮区熔法 floating zone melting
悬键 dangling bond
悬链线 catenary
悬桥 suspension bridge
悬置 suspension
旋磁比 gyromagnetic ratio
旋磁器件 gyromagnetic device
旋磁效应 gyromagnetic effect
旋电效应 gyroelectric effect
旋度 curl；rot；rotation
旋度场 curl field
旋光 rotation
旋光度 optical rotation
旋光改变作用 mutarotation
旋光计 polarimeter
旋光率 specific rotation
旋光色散 ORD (optical rotary dispersion)
旋光性 optical activity；optical rotation
旋光仪 azimuth polarimeter
旋光异构体 optical isomer
旋节线 spinodal curve
旋节线分解 spinodal decomposition
旋节线分离 spinodal decomposition
旋进 precession
旋镜 rotating mirror
旋量 spinor
旋流 rotational flow
旋涡偶 vortex pair
旋涡星系 spiral galaxy
旋涡轴 vortex axis
旋转泵 rotary pump
旋转变换 rotational transform
旋转变流机 rotary converter
旋转镜 rotating mirror
旋转罗盘 gyrocompass
旋转轴 axis of rotation
旋子 roton
选模 mode selection
选频放大器 selective amplifier
选择定则 selection rule
选择反射 selective reflection
选择激发 selective excitation
薛定谔方程 Schrödinger equation
学说 theory
雪崩 avalanche
雪崩击穿 avalanche breakdown
寻常光线 ordinary ray
寻常射线 ordinary ray
寻星镜 finder
巡游电子 itinerant electron
循环 circle；cycle；loop；revolution
循环量子 fluxoid
讯号 signal

Yy

压差阻力　pressure drag
压磁效应　piezomagnetic effect
压电换能器　piezoelectric transducer
压电晶体　piezocrystal；piezoelectric crystal
压电石英　piezoelectric quartz；piezoid
压电陶瓷　piezoelectric ceramics
压电体　piezoelectrics
压电效应　piezoelectric effect
压电振荡器　piezoelectric oscillator
压电振子　piezoelectric vibrator
压光效应　piezo-optic effect
压降　pressure drop
压力　pressure
压力表　pressure meter
压力补偿器　pressure compensator
压力差　pressure difference
压力场　pressure field
压力传感器　pressure transducer
压力传送器　pressure transmitter
压力方程　pressure equation
压力计　pressure gauge；pressure meter
压力开关　pressure switch
压力扩散　pressure diffusion
压力曲线　pressure curve
压力系数　pressure coefficient
压力效应　pressure effect
压强　pressure
压强差　pressure difference
压强传送器　pressure transmitter
压强计　piezometer；pressure gauge
压强开关　pressure switch
压强扩散　pressure diffusion
压强系数　pressure coefficient
压球硬度　Brinell hardness
压水反应堆　PWR（pressurized-water reactor）
压缩　compression
压缩比　compression ratio
压缩波　compressional wave
压缩率　compressibility；compression ratio
压缩式真空计　McLeod gauge
压温相图　pressure-temperature phase diagram
压胁强　pressure stress
压应力　pressure stress
压致电阻效应　piezoresistive effect
压致频移　pressure shift
压致损失　pressure drop
压致移动　pressure shift
压阻效应　piezoresistive effect
雅可比行列式　Jacobian（determinant）
亚毫米波　submillimetre wave
亚级　substate
亚临界的　subcritical
亚能级　substate
亚声速　subsonic speed
亚态　substate
亚铁磁性　ferrimagnetism
亚铁电体　ferrielectrics
亚稳定性　metastability
亚稳界线　spinodal curve
亚稳平衡　metastable equilibrium
亚稳态　metastable state；semistable state

亚稳相 metastable phase
亚原子粒子 subatomic particle
氩离子激光器 argon ion laser
氩离子刻蚀 argon ion etching
烟 smoke
湮没 annihilation
湮没辐射 annihilation radiation
延迟 retardation
延迟板 retardation plate
延迟线 delay line
延脆转变 ductile-brittle transition
延度 ductility
延伸率 extensibility
延伸性 expansivity; extensibility
延时摄影 time-lapse photography
延性 ductility
延性材料 ductile material
延性断裂 ductile fracture
严格解模型 exactly solved model
研磨料 abrasive
颜色 colour
衍射 diffraction
衍射波 diffracted wave
衍射光栅 diffraction grating
衍射光学 diffraction optics
衍射级 order of diffraction
衍射计 diffractometer
衍射角 angle of diffraction; diffraction angle
衍射屏 diffraction screen
衍射条纹 diffraction fringe
衍射图样 diffraction pattern
衍射仪 diffractometer
衍生物 derivative
掩 occultation
掩蔽 mask
掩模 mask
掩星 occultation
眼点距 eye relief
眼睛 eye
眼睛间隙 eye relief
眼科光学 ophthalmic optics
演化 evolution
演化规则 evolution rule
演化算符 evolution operator
演算 calculus
验磁器 magnetoscope
验电器 electroscope
焰熔法 flame fusion method
赝标量 pseudoscalar
赝矢量 pseudovector
赝势 pseudopotential
赝势法 method of pseudopotential
赝张量 pseudotensor
扬声器 loudspeaker; speaker
阳电荷 positive charge
阳电性的 electropositive
阳电子 positron
阳极 anode; positive pole
阳极射线 canal ray
阳离子 cation; positive ion
杨氏双缝 Young's slits
杨氏双缝实验 Young's double-slit experiment
杨氏弹性模量 Young modulus of elasticity
杨(振宁)-米尔斯场 Yang-Mills field
杨(振宁)-米尔斯规范场 Yang-Mills field
佯谬 paradox
氧化氘 deuterium oxide; heavy water
氧化物半导体 oxide semiconductor
样本方差 sample variance
样本函数 sample function
样本空间 sample space
样本值 sample value
幺模变换 unimodular transformation
幺旋 unitary spin
幺正变换 unitary transformation
幺正矩阵 unitary matrix
幺正群 unitary group
幺正算符 unitary operator
幺正性 unitarity
摇溶现象 thixotropy

遥测 remote sensing
遥测光度计 telephotometer
遥测温度计 telethermometer
遥测仪 telegauge；telemeter
遥感 remote sensing
遥控 remote control；telecontrol
遥控机械学 telemechanics
遥控键 teleswitch
遥控开关 teleswitch
遥控力学 telemechanics
耀度 brilliance
冶金术 metallurgy
冶金学 metallurgy
叶轮风速计 vane anemometer
叶片 wheel
曳力系数 drag coefficient
曳引系数 drag coefficient
夜间辐射 nocturnal radiation
夜视 night vision
液槽气压计 cistern barometer
液氮 liquid nitrogen
液滴模型 liquid-drop model
液氦 liquid helium
液化 deliquescence；fluidization；liquefaction
液化点 liquefaction point
液化器 liquefier
液化石油气 LPG（liquefied petroleum gas）
液晶 liquid crystal
液晶显示器 LCD（liquid crystal display）
液晶相 liquid crystal phase
液体 liquid
液体比重计 hydrometer
液体激光器 liquid laser
液相外延生长 rheotaxial growth
液相线 liquidus（line）
液性 fluidity
一贯制单位 coherent units
一级光谱 spectrum of first order
一级相变 first-order phase transition
依数性 colligative properties
仪器 instrument；set
仪器光学 instrumental optics
移动 motion
移动速率 travelling speed
移后 lag
移相器 phaser；phase shifter
以太 ether
以太风 ether wind
以太漂移 ether drift
以太曳引 ether drag
钇 Y（yttrium）
钇铝石榴石激光器 YAG laser
异常 anomaly
异极电荷 heterocharge
异相 out of phase
异质结 heterojunction
异质结构 heterostructure
异质结激光器 heterojunction laser；heterolaser
异质结结构 heterostructure
异质外延 heteroepitaxy
抑制栅极 suppressor grid
译码器 translator
易磁化方向 easy direction for magnetization
易磁化轴 easy axis of magnetization
易熔合金 fusible alloy
逸出功 work function
逸度 fugacity
逸性 fugacity
翼剖面 airfoil profile；airfoil section
翼型 aerofoil；airfoil（profile）
镱 Yb（ytterbium）
因次分析 dimensional analysis
因果关系 causality
因果律 causality
因果性 causality
因式分解 factorization
因数 factor
因素 factor
因子 factor
因子分解 factorization
因子分解条件 factorization condition
因子化 factorization

阴电的 electronegative
阴电荷 negative charge
阴电性 electronegativity
阴电子 negatron
阴极 cathode；negative pole
阴极板 negative plate
阴极保护 cathodic protection
阴极防蚀 cathodic protection
阴极射线 cathode ray
阴极射线发光 cathodoluminescence
阴极射线管 CRT（cathode-ray tube）
阴极射线示波器 CRO（cathode-ray oscilloscope）
阴介子 negative meson
阴离子 anion；negative ion
阴历月 synodic month
阴影 shadow
阴影区 shadow region
阴影效应 shadow effect
阴质子 negative proton
音叉 tuning fork
音调 pitch；tone
音符 note
音阶 （musical）scale
音量 loudness；volume
音频 audiofrequency
音品 quality of sound；timbre
音色 quality of sound；timbre
音障 sound barrier
音质 quality of sound；timbre
殷钢 Invar
银河系 Galaxy；Milky Way（Galaxy）
银河星团 galactic cluster
引出线 output
引力 attractive force；gravitation；gravitational force
引力半径 gravitational radius
引力波 gravitational wave
引力常量 gravitational constant
引力场 gravitational field
引力场强度 gravitational field strength
引力辐射 gravitational radiation
引力红移 Einstein（red）shift；gravitational redshift
引力凝聚 gravitational condensation
引力时间延缓 gravitation time dilation
引力势 gravitational potential
引力坍缩 gravitational collapse
引力透镜 gravitational lens
引力微子 gravitino
引力位能 gravitational potential energy
引力相互作用 gravitational interaction
引力移位 gravitational shift
引力质量 gravitational mass
引力中心 barycentre
引力子 graviton
引擎 engine
引燃 ignition
引伸计 extensometer
引向天线 Yagi aerial
隐蔽 occultation
隐变量 hidden variable
隐藏 occultation
隐对称性 hidden symmetry
隐灭常数 decay constant
印刷电路 printed circuit
印刷电路板 PCB（printed circuit board）
英尺 foot
英寸 inch
英国单位制 British system of units
英国热单位 Btu（British thermal unit）
英里 mile
英两 oz（ounce）
英制 English system
英制单位 Imperial units
迎角 incidence angle
荧光 fluorescence；fluorescent light
荧光灯 fluorescent lamp；fluorescent light
荧光分析 fluorescence analysis

荧光检测 fluoroscopy
荧光镜 fluoroscope
荧光屏 screen；telescreen
荧光屏照片 photofluorogram
荧光摄影术 photofluorography
荧光学 fluoroscopy
荧光照相 photofluorography
蝇眼透镜 fly's-eye lens
影 shadow
应变 strain
应变规 strain gauge
应变计 strain gauge
应变硬化 strain hardening
应答器 transponder
应力 stress
应力双折射 stress birefringence
应力硬化 stress hardening
应力张量 stress tensor
应用光学 applied optics
应用力学 applied mechanics
应用软件 applications software
应用物理(学) applied physics
映像 image
硬磁材料 hard magnetic material
硬磁泡 hard magnetic bubble
硬度 hardness
硬光子 hard proton
硬过程 hard process
硬激发 hard excitation
硬件 hardware
硬模 hard mode
硬性辐射 hard radiation
永磁场 permanent magnetic field
永磁发电机 magneto
永磁体 permanent magnet
永磁性 permanent magnetism
永电体 electret
永动机 perpetual motion machine
永恒运动 perpetual motion
永久气体 permanent gas
用电安全 electric safety
优值 Q-value
油浸电容器 oil capacitor；oil condenser
油浸透镜 oil immersion lens
油浸物镜 oil immersion objective
油扩散泵 oil diffusion pump
油雾 oil vapour
油液减震 oil damping
铀 U（uranium）
铀后元素 transuranic element
铀系 uranium series
游标(尺) vernier
游标卡尺 vernier caliper
有辐射的 radioactive
有功电流 active current
有功电压 active voltage
有功负荷 active load
有功功率 active power
有功元件 active element
有极分子 polar molecule
有理数 rational number
有穷数列 finite series
有声电影 teletalking
有限级数 finite series
有限转动 finite rotation
有像差光学部件 aberrated optics
有像差光学系统 aberrated optics
有像差透镜系统 aberrated lens system
有效半径 effective radius
有效电导率 effective conductivity
有效电荷 effective charge
有效电流 active current
有效电压 active voltage
有效范围 effective range
有效功率 active power
有效截面 effective cross-section
有效距离 effective range
有效孔径 effective aperture
有效口径 effective aperture
有效力程 effective range
有效量程 effective range
有效面积 active area
有效势 effective potential

有效数字 significant figures
有效温度 effective temperature
有效线宽 effective line width
有效值 effective value
有效质量 active mass; effective mass
有心力 central force
有心力场 central field
有序 order
有序无序转变 order-disorder transition
有旋电场 curl electric field
有源成像系统 active imaging system
有源电路 alive circuit
有源负载 active load
有源器件 active device
有源腔 active cavity
有源元件 active element
酉变换 unitary transformation
酉矩阵 unitary matrix
酉群 unitary group
酉算子 unitary operator
右矢 ket
右手定则 right-hand rule
右手流 right-handed current
右手螺旋定则 right-handed screw rule
右旋晶体 right-handed crystal
右旋性的 dextrorotatory
右(旋)圆偏振 right-hand circular polarization
诱发裂变 induced fission
余对数 cologarithm
余辉 afterglow
余留像 residual image
余热 afterheat
余弦定律 cosine rule
余像 afterimage; residual image
隅角棱镜 corner cube
逾力 excess force
逾流 excess flux
逾熵 excess entropy
逾熵产生 excess entropy production
逾熵平衡方程 excess entropy balance equation
与非电路 NAND circuit
与非门 AND-NOT gate
与或非门 AND-OR-INVERT gate
与或门 AND-OR gate
与门 AND gate
宇称 parity
宇称不守恒 parity nonconservation
宇称检验 parity check
宇称守恒 parity conservation
宇称守恒定律 law of conservation of parity
宇称(性)〖波(动)函数的〗 parity of wave function
宇航学 astronautics
宇宙 cosmos; universe
宇宙背景辐射 cosmic background radiation
宇宙背景探测器 COBE (cosmic background explorer)
宇宙常数 cosmological constant
宇宙尘 cosmic dust; meteoroid
宇宙丰度 cosmic abundance
宇宙辐射 cosmic radiation
宇宙光 cosmic ray
宇宙监督假设 cosmic censorship hypothesis
宇宙空间 aerospace
宇宙论 cosmology
宇宙年龄 age of the universe
宇宙膨胀 expansion of the universe
宇宙物理学 cosmophysics
宇宙线 cosmic ray
宇宙学 cosmology
宇宙学红移 cosmological redshift
宇宙学原理 cosmological principle
宇宙噪声 cosmic noise
宇宙中微子 cosmic neutrino
预处理 preprocessing
预电离 preionization
预热 preheating

预注入器 preinjector
域 domain
阈 threshold
阈能 threshold energy
阈频率 threshold frequency
阈值功率 threshold power
阈值条件 threshold condition
元胞法 cellular method
元胞自旋 cell spin
元电荷 elementary charge
元功 elementary work
元光束 elementary beam
元激发 elementary excitation
元激发谱 elementary excitation spectrum
元件 element
元素 element
元素周期表 periodic table of elements
原胞 primitive cell
原点 origin
原电池 galvanic cell; primary cell; voltaic cell
原恒星 protostar
原能 proper energy
原绕组 primary winding
原色 primary colour
原始数据 raw data
原像 preimage; primary image
原星系 protogalaxy
原子 atom
原子尘 radioactive fallout
原子弹 atomic bomb
原子反应堆 atomic pile
原子辐射 atomic radiation
原子光谱 atomic spectrum
原子轨道函数 atomic orbital
原子核 (atomic) nucleus
原子核物理学 (atomic) nuclear physics; nucleonics
原子化热 heat of atomization
原子价 valence; valency
原子间力 interatomic force
原子结构 atomic structure
原子科学 atomics
原子力显微镜 AFM (atomic force microscope)
原子量 atomic weight
原子论 atomic theory
原子模型 atomic model
原子能 atomic energy; nuclear energy
原子碰撞 atomic collision
原子谱线 atomic spectral line
原子散射因子 atomic scattering factor
原子束 atomic beam
原子体积 atomic volume
原子物理学 atomic physics
原子吸收光谱学 atomic absorption spectroscopy
原子序数 atomic number; proton number
原子质量 atomic mass
原子质量单位 a.m.u. (atomic mass unit)
原子钟 atomic clock
原子自旋 atomic spin
圆 circle
圆规 compass
圆环域 annulus
圆孔衍射 circular hole diffraction
圆盘衍射 circular disk diffraction
圆偏振 circular polarization
圆频率 angular frequency; circular frequency; pulsatance
圆双折射 circular birefringence
圆周率 π (pi)
圆周运动 circular motion
圆柱调和函数 cylindrical harmonics
圆锥发射 cone emission
圆锥体 cone
圆锥效应 cone effect
源(点) source
远场 far field
远场条件 far field condition
远场图形 far field pattern
远场图样 far field pattern
远地点 apogee

远(拱)点 apoapsis
远红外(线) far infrared
远距分光镜 telespectroscope
远距光度计 telephotometer
远日点 aphelion
远摄镜头 telelens；telephoto lens
远摄照相 telephoto
远视 far sight；hyper(metr)opia；long-sightedness
远视眼 presbyopia
远心的 telecentric
远心光阑 telecentric stop
远心光束 telecentric beam
远心系统 telecentric system
远月点 apocynthion；apolune
远震 teleseism
远紫外激光器 uvaser
约化质量 reduced mass
约瑟夫森器件 Josephson device
约瑟夫森效应 Josephson effect
约束 constraint
约束力 constraining force
约束时间 confinement time
约束运动 constrained motion
月亮 moon
月偏食 partial lunar eclipse
月球 moon
月球学 selenology
月食 lunar eclipse
月相 phases of the moon
跃迁 jump；transition
跃迁概率 transition probability
跃迁几率 transition probability
跃迁矩阵 transition matrix
跃迁能 transition energy
越程 skip distance
越距 skip distance
云室 cloud chamber
云纹图 moiré pattern
匀加速度 uniform acceleration
匀速 uniform speed；uniform velocity
匀速圆周运动 uniform circular motion
匀速运动 uniform motion
陨石 aerolite
陨星 meteorite
运动 motion
运动方程 equation of motion
运动黏度 kinematic viscosity
运动学 kinematics
运动学方程 kinematic equation
运流 convection
运流电流 convection current
运送电子 convey electron
运算放大器 op-amp (operational amplifier)
运载火箭 carrier rocket
运载气体 carrier gas
晕 halo
晕环 corona

Zz

匝数比　turns ratio
杂光像　spurious image
杂乱散射　incoherent scattering
杂散电流　stray current
杂散电容　stray capacitance; stray capacity
杂散辐射　stray radiation
杂色光　heterochromatic light; heterogeneous light
杂像　parasitic image
杂质　impurity
杂质电阻率　impurity resistivity
杂质能级　impurity level
灾变　catastrophe
灾变性激光损伤阈　catastrophic laser-damage threshold
载波　carrier (wave)
载波频率　carrier-wave frequency
载荷　load
载荷子　charge carrier
载流子　(charge) carrier
载流子复合　carrier recombination
载流子密度　carrier density
载流子浓度　carrier concentration
载流子迁移率　carrier mobility
载流子寿命　carrier lifetime
载流子输运　carrier transport
载流子注入　carrier injection
载体　carrier
再辉　recalescence
再凝　regelation
再生能源　renewable energy source
再现性　reproducibility
暂(时)磁性　temporary magnetism
暂态　transience; transient state
暂态平衡　transient equilibrium
暂态运动　transient motion
早期宇宙　early universe
噪扰　noise
噪扰源　noise source
噪声　noise
噪声背景　noise background
噪声波形图　noise audiogram
噪声测试器　noise meter
噪声二极管　noise diode
噪声幅度　noise amplitude
噪声函数　noise function
噪声环境　noise environment
噪声计　noise meter
噪声剂量　noise dose
噪声剂量计　noise dose meter
噪声降低　noise abatement; noise reduction
噪声滤波器　noise filter
噪声脉冲　noise pulse
噪声能量　noise energy
噪声容限　noise margin
噪声输出　noise output
噪声污染　noise pollution
噪声系数　noise factor; noise figure
噪声陷波器　noise trap
噪声消除器　noise killer
噪声抑制　noise abatement; noise suppression
噪声抑制器　noise suppressor
噪声因数　noise factor; noise figure
噪声源　noise source

噪音 noise
泽贝克效应 Seebeck effect
泽格测温锥 pyrometric cone; Seger cone
增湿 humidification
增湿器 humidifier
增透膜 antireflecting film
增益 gain
增益饱和 gain saturation
增益系数 gain coefficient
增原子 adatom
增殖 proliferation
增殖材料 fertile material
增殖反应堆 breeder reactor; fast breeder
增殖性物质 fertile material
增殖因数 multiplication factor
闸流管 thyratron
闸流晶体管 thyristor
窄带 narrow-band
窄脉冲 narrow pulse
斩光器 light chopper
张弛振荡器 relaxation oscillator
张力 tensile force; tensility; tension
张力测力计 tension dynamometer
张力计 extensometer; tensiometer; tensometer
张力扩散 tensodiffusion
张力数 tensile figure
张量 tensor
张量变换 tensor transformation
张量分析 tensor analysis
张量函数 tensor function
张量密度 tensor density
张量算子 tensor operator
张丝 tension spring
张胁变 tensile strain
张胁强 tensile stress
张应变 tensile strain
张应力 tensile stress
章动 nutation
涨落 fluctuation
涨落场论 fluctuation field theory
涨落耗散定理 fluctuation-dissipation theorem
涨落力 fluctuating force
涨落项 fluctuation term
障碍物 obstacle
兆赫 MHz (megahertz)
兆欧计 megohmmeter
照度 illuminance; illumination
照度计 illuminometer; luminometer; luxmeter
照明 illumination
照明装置 lighting device
照片 photograph
照射量 exposure
照相机 camera
照相记时仪 photochronograph
照相镜头 photographic lens
照相密度 photographic density
照相排版 photocomposition
照相术 photography
照相透镜 photographic lens
照相望远镜 photographic telescope
照相制版 photoengraving
遮断效应 blocking effect
遮光器 light chopper
折叠谱 folded spectrum
折叠腔 folded cavity
折叠(腔)激光器 folded laser
折叠势 folded potential
折光常数 refractive constant
折光度 dioptre
折光率 refractive index
折合质量 reduced mass
折射 refraction
折射本领 refractivity
折射常数 refractive constant
折射定律 law of refraction; refraction law
折射光学 dioptrics
折射计 refractometer
折射角 angle of refraction; refraction angle
折射率 refractive index; refractivity
折射器 refractor

折射式望远镜 refracting telescope；refractor
折射系数 refraction coefficient
折射线 refracted ray
折射性 refractivity
折轴望远镜 coudé telescope
锗 Ge (germanium)
针孔成像 pinhole imaging
针孔照相机 pinhole camera
真近点角 true anomaly
真空 vacuum
真空泵 vacuum pump
真空测量 vacuum measurement
真空场 vacuum field
真空萃取 vacuum extraction
真空灯 vacuum lamp
真空点火 vacuum firing
真空度指示器 vacuum indicator
真空镀膜 vacuum coating
真空断流器 vacuum breaker；vacuum circuit-breaker
真空断续器 vacuum interrupter
真空阀 vacuum valve
真空放电 vacuum discharge
真空管 vacuum tube；(vacuum) valve
真空光电管 vacuum photocell；vacuum phototube
真空规 vacuum gauge
真空盒 vacuum box
真空机械手 vacuum tweezer
真空计 vacuometer；vacuscope；vacuum gauge
真空介电常量 dielectric constant of vacuum
真空阱 vacuum trap
真空净化 vacuum purification
真空聚集槽 vacuum trap
真空绝缘 vacuum insulation
真空开关 vacuum switch
真空空间 vacuum space
真空冷却 vacuum cooling
真空镊子 vacuum tweezer
真空膨胀计 vacuum dilatometer
真空膨胀器 vacuum expander
真空瓶 vacuum flask；vacuum vessel
真空钳 vacuum tweezer
真空清洁器 vacuum cleaner
真空热电堆 vacuum thermopile
真空热电偶 vacuum thermocouple
真空容器 vacuum vessel
真空熔炼 vacuum melting
真空室 vacuum chamber；vacuum space
真空损失 vacuum leak
真空态 vacuum state
真空提纯 vacuum purification
真空提取 vacuum extraction
真空涂敷 vacuum coating
真空温差电堆 vacuum thermopile
真空温差电偶 vacuum thermocouple
真空物理学 vacuum physics
真空吸尘器 vacuum cleaner
真空系数 vacuum coefficient
真空泄漏 vacuum leak
真空压力计 vacuum manometer
真空仪 vacuscope
真空增强器 vacuum intensifier
真空闸 vacuum brake
真空长成晶体 vacuum grown crystal
真空蒸馏 vacuum distillation
真空蒸馏塔 vacuum tower
真空整流管 vacuum valve
真空制动 vacuum brake
真空装置 vacuum plant
真倾角 angle of dip
真实气体 real gas
真位置 true place
真误差 true error
真值 actual value；true value
真值表 truth table
振荡 oscillation；vibration
振荡电流 oscillating current
振荡晶体 oscillating crystal
振荡偶极子 oscillating dipole
振荡器 oscillator
振荡宇宙 oscillating universe

振动 oscillation；vibration
振动激发 vibrational excitation
振动激励 vibrational excitation
振动模(式) mode of vibration
振动能带 vibrational band
振动谱带 vibrational band
振幅 amplitude
振幅分割 division of amplitude
振幅畸变 amplitude distortion
振幅调制 AM (amplitude modulation)
振幅型光栅 amplitude grating
振子 oscillator
震深 depth of focus
震源 epicentre
震中 epicentre
镇流电阻器 ballast resistor
蒸镀 deposition；evaporation
蒸发 evaporation
蒸发模型 evaporation model
蒸发谱 evaporation spectrum
蒸发热 heat of evaporation
蒸汽 vapour
蒸汽泵 vapour pump
蒸汽机 steam engine
蒸汽量热器 steam calorimeter
蒸汽密度 vapour density
蒸汽(涡)轮机 steam turbine
蒸汽吸水机 vacuum pump
蒸汽相 vapour phase
蒸汽压 vapour pressure；vapour tension
蒸汽压降低 lowering of vapour pressure
蒸汽蒸馏 steam distillation
整流 rectification
整流器 rectifier
正比计数器 proportional counter
正常色散 normal dispersion
正常态 normal state
正电荷 positive charge
正电性的 electropositive
正电子 antielectron；positron
正电子发射层析术 PET (positron emission tomography)
正电子素 positronium
正反共轭 charge conjugation
正反共轭宇称 C-parity
正反馈 positive feedback
正负电子对产生 electron-positron pair creation
正负电子对撞机 electron-positron collider
正极 anode；positive pole
正极板 positive plate
正极性 positive polarity
正交的 orthorhombic
正交函数 orthogonal function
正交化 orthogonalization
正交加速度 normal acceleration
正交棱镜 cross prisms
正交力 normal force
正交偏振器 crossed polarizers
正交性 orthogonality
正交状态 orthogonality
正交轴 orthogonal axis
正晶体 positive crystal
正离子 cation；positive ion
正碰 direct impact；head-on collision
正切 tangent
正切电流计 tangent galvanometer
正氢 orthohydrogen
正入射 normal incidence；vertical incidence
正态分布 Gaussian distribution；normal distribution
正弦波 sine wave；sinusoidal wave
正弦定律 sine rule
正弦曲线 sine curve；sinusoid
正弦振荡器 sinusoidal oscillator
正像 erect image
正像棱镜 erecting prism
正胁强 normal stress
正应力 normal stress
正则变换 canonical transformation
正则方程 canonical equation
正则分布 canonical distribution

正则坐标 canonical coordinates
正子 antielectron
支导线 open traverse
支点 foot；fulcrum；pivot
支壳层 subshell
支路 branch
支轴 fulcrum
枝(蔓)晶(体)生长 dendritic growth
织构 texture
脂状相 smectic phase
直角坐标 Cartesian coordinates
直接带隙 direct band gap
直接关联 direct correlation
直接吸收 direct absorption
直流(电) DC；d. c.（direct current）
直流电动发电机 dynamotor
直流电动机 direct current motor
直流电桥 direct current bridge
直流马达 direct current motor
直射变换 collinear transformation；collineation
直视棱镜 direct vision prism
直线传播 rectilinear propagation
直线电动机 linear motor
直线箍缩 linear pinch
直线加速器 linac；linear accelerator
直线运动 rectilinear motion
植入 implantation
只读存储器 ROM（read-only memory）
纸带 paper tape
指南针 compass；magnetic needle
指示器 pointer
指数 exponent
指数定理 index theorem
指数增长 exponential growth
指针 pointer
制导激光束 guiding laser beam
制动火箭 retrorocket
制动距离 braking distance；stopping distance
制动器 decelerator
制动闸 brake
制镜合金 speculum
制冷 refrigeration
制冷机 refrigerator
制冷剂 refrigerant
制冷器 refrigerator
制约 constraint
质点 mass point；material point；particle
质点动力学 particle dynamics
质点速度 particle velocity
质量 mass
质量不灭 conservation of mass
质量公式 mass formula
质量过剩 mass excess
质量减损 mass decrement
质量亏损 mass defect
质量密集物质 mascon
质量守恒 conservation of mass
质量守恒定律 law of conservation of mass
质量数 mass number
质量数守恒 mass-number conservation
质量盈余 mass excess
质量转移 mass transfer
质量阻止本领 mass stopping power
质量作用 mass action
质量作用(定)律 mass action law
质能等价性 mass-energy equivalence
质能关系 mass-energy relation
质谱 mass spectrum
质谱分析器 mass spectrometer
质谱仪 mass spectrometer
质心 barycentre；centre of mass；centroid
质心(参考)系 CMS（centre-of-mass system）
质子 proton
质子共振 proton resonance
质子偶素 protonium
质子数 proton number
质子衰变 proton decay

质子同步加速器 proton synchrotron
质子自旋共振 proton spin resonance
致断胁强 breaking stress
致断应力 breaking stress
致密天体 compact object
致密星系 compact galaxy
致命杂质 deathnium
致偏电极 deflecting electrode
滞后 hysteresis；lag
滞后回线 hysteresis loop
滞后曲线 hysteresis curve
滞后效应 hysteresis
滞后作用 hysteresis
置闰 intercalation
置信度 CL（confidence level）
置信区间 confidence interval
置信水平 CL（confidence level）
中波 mw（medium wave）
中等耦合 intermediate coupling
中国物理学会 Chinese Physical Society
中和热 heat of neutralization
中继 relay
中间矢量玻色子 IVB（intermediate vector boson）
中间态 intermediate state
中间图像 intermediate image
中间相 intermediate phase
中阶梯光栅 echelle grating
中介相 mesophase
中能中子 intermediate neutron
中频 IF（intermediate frequency）；MF（medium frequency）
中数 mean
中微子 neutrino
中线 median
中心对称 central symmetry
中性的 neutral
中性粒子 chargeless particle
中性流 NC（neutral current）
中性平衡 neutral equilibrium
中性原子 neutral atom
中央处理单元 CPU（central processing unit）
中央极大 central maximum
中央条纹 central fringe
中值 median
中质差 isotopic number
中子 neutron
中子弹 neutron bomb
中子发生器 neutron generator
中子反应 neutron reaction
中子俘获 neutron death
中子辐照 neutron irradiation
中子光学 neutron optics
中子过剩 neutron excess
中子活化分析 NAA（neutron activation analysis）
中子扩散 neutron diffusion
中子能谱 neutron spectrum
中子平均自由程 neutron mean free path
中子谱学 neutron spectroscopy
中子谱仪 neutron spectrometer
中子散射 neutron scattering
中子摄谱仪 neutron spectrograph
中子数 neutron number
中子探测器 neutron detector
中子通量 neutron flux
中子温度 neutron temperature
中子吸收剂 neutron sponge
中子泄漏 neutron leakage
中子星 neutron star
中子学 neutronics
中子循环 neutron cycle
中子衍射 neutron diffraction
中子仪器 neutron instrument
中子源 neutron source
中子增殖 neutron multiplication
中子质子数差 isotopic number
中子总数 neutron inventory
终点弹道学 terminal ballistics
终端条件 terminal condition
终端阻抗 terminal impedance
终极速度 terminal speed；terminal velocity
终极速率 terminal speed；terminal velocity

终态 final state
钟(青)铜 bell metal
重垂线 plumb line
重费米子 heavy fermion
重光子 heavy photon
重核 heavy nucleus
重核子 hyperon
重夸克 heavy quark
重离子 heavy ion
重离子反应 heavy-ion reaction
重离子放射性 heavy-ion radioactivity
重离子碰撞 heavy-ion collision
重力 gravitational force; gravity
重力场 gravitational field; gravity field
重力加速度 acceleration of gravity
重力漂移 gravitational drift
重力势 gravitational potential
重量 weight
重轻子 heavy lepton
重氢 deuterium; heavy hydrogen
重氢核 deuteron
重氢化合物 deuterated compound
重水 deuterium oxide; heavy water
重心 barycentre; centre of gravity
重子 barion; baryon
重子数 baryon number
周 cycle
周率 frequency
周期 circle; cycle; period
周期律 periodic law
周期性 periodicity
周期运动 periodic motion
周转圆 epicycle
轴 axis; shaft
轴封环 packing ring
轴矢(量) axial vector
轴外光线 oblique ray
轴外像差 off-axis aberration
轴向对称 axial symmetry
轴向放大率 axial magnification; longitudinal magnification
轴向加速度 axial acceleration
轴向力 axial force
轴向量 axial vector
轴向像差 axial aberration
轴子 axion
珠光体 pearlite
逐区精炼 zone refining
烛光 candle power
主波 principal wave
主程序 master routine
主导极点 leading pole
主导粒子 leading particle
主导项 leading term
主点 principal point
主动输运 active transport
主方程 master equation
主观亮度 subjective brightness
主光线 principal ray
主光轴 primary optical axis
主极大 principal maximum
主焦点 prime focus; principal focus
主镜 primary mirror
主量子数 principal quantum number
主面 principal plane
主频(率) basic frequency
主矢(量) principal vector
主速度 principal velocity
主线系 principal series
主向量 principal vector
主胁强 principal stress
主序星 main-sequence star
主应力 principal stress
主折射率 principal refractive index
主轴 main shaft; principal axis
主转动惯量 principal moment of inertia
注入式激光器 injection laser
驻波 standing wave; stationary wave
驻极体 electret
驻粒子 stationary particle

柱面波 cylindrical wave
柱面极坐标 cylindrical polar coordinates
柱面镜 cylindrical mirror
柱面透镜 cylindrical lens
柱塞 plunger
柱谐函数 cylindrical harmonics
铸铁 cast iron
转变 transition
转变层 transition layer
转变点 transition point
转变能 transition energy
转变损失 transition loss
转变温度 transition temperature
转导 transduction
转换电子 conversion electron
转换反应堆 converter reactor
转换效率 conversion efficiency
转换因子 conversion factor
转接器 adapter；commutator
转向目镜 cranked eyepiece
转像棱镜 inverting prism
转移常数 transfer constant
转移函数 transfer function
转移恒量 transfer constant
转移矩阵 transfer matrix
转移曲线 transfer curve
转移系数 transfer coefficient；transfer factor
转移因子 transfer factor
转移阻抗 transfer impedance
转动 rotation
转动惯量 moment of inertia
转动即时轴 instantaneous axis of rotation
转动角 angle of rotation
转动能 rotational energy
转动瞬轴 instantaneous axis of rotation
转动体 rotor
转动轴 axis of rotation；rotation axis
转矩 torque；torsion；twisting moment
转数 rotation number
转速计 tachometer；turnmeter
转子 rotor
装置 device；equipment；set
撞击中心 centre of percussion
追踪器 tracker
锥体 cone
锥形物 cone
准费米能级 quasi-Fermi level
准分子 quasi-molecule
准分子激光器 excimer laser；exciplex laser
准晶 quasicrystal
准静态过程 quasi-static process
准粒子 quasi-particle
准裂变 quasi-fission
准平衡 quasi-equilibrium
准确度 accuracy
准线 directrix
准线性方程 quasilinear equation
准直 collimation
准直管 collimator
准直光束 collimated beam
准直器 aligner；collimator
灼热剂 thermite
姿态角 attitude angle
子核 daughter nucleus
子午线 meridian
子相 product phase
紫外波段 ultraviolet band
紫外的 UV（ultraviolet）
紫外辐射 ultraviolet radiation
紫外光谱图 ultraviolet spectrogram
紫外光子 ultraviolet photon
紫外谱带 ultraviolet band
紫外区 ultraviolet band
紫外摄谱仪 ultraviolet spectrograph
紫外线 ultraviolet ray；UV（ultraviolet）
紫外线灯 ultraviolet lamp
紫外线辐照 ultraviolet irradiation
紫外线激光器 UV laser
紫外线剂量计 ultraviolet dosimeter
紫外线探测器 ultraviolet

detector
紫外线显微镜 ultraviolet microscope
紫外灾难 ultraviolet catastrophe
字 word
字节 byte
自变数 argument
自掺杂 autodoping
自动射线照相术 autoradiography
自动同步机 teletorque
自发磁化(强度) spontaneous magnetization
自发发射 spontaneous emission
自发(核)分裂 spontaneous fission
自发裂变 spontaneous fission
自发跃迁 spontaneous transition
自感 self-inductance; self-induction
自感系数 self-inductance
自感应 self-induction
自激振荡 self-excited oscillation
自激振动 autovibration; self-excited oscillation
自能 self-energy
自耦变压器 autotransformer
自洽场 self-consistent field
自洽解 self-consistent solution
自洽性 self-consistency
自然波长 natural wavelength
自然单位 natural unit
自然电流 natural current
自然对数 Napierian logarithm; natural logarithm
自然丰度 natural abundance
自然光 natural light
自然晶面 natural crystal plane
自然哲学 natural philosophy
自然振动 natural vibration
自然阻尼 natural damping
自然坐标 natural coordinates
自燃 pyrophoricity
自适应光学 adaptive optics
自适应光学系统 adaptive optical system; adaptive optics
自行 proper motion
自旋 spin
自旋波 spin wave
自旋波共振 spin wave resonance
自旋玻璃 spin glass
自旋共振 spin resonance
自旋轨道分裂 spin-orbit splitting
自旋轨道耦合 spin-orbit coupling
自旋简并性 spin degeneracy
自旋角动量 spin angular momentum
自旋量子数 spin quantum number
自旋密度 spin density
自旋能 spin energy
自旋效应 spin effect
自旋张量 spin tensor
自由程 free path
自由磁荷 free magnetic charge
自由电荷 free charge
自由电子 free electron
自由电子激光器 free-electron laser
自由电子模型 free-electron model
自由度 degree of freedom
自由焓 free enthalpy
自由焓判据 free enthalpy criterion
自由降落 free fall
自由空间 free space
自由离子模型 free-ion model
自由落体加速度 acceleration of free fall
自由能 free energy
自由能判据 free energy criterion
自由膨胀 free expansion
自由矢量 free vector
自由下落 free fall
自转 rotation
自转角 angle of rotation
自准光谱仪 autocollimatic spectroscope

自准直 autocollimation
自组织 self-organization
综合孔径 synthetic aperture
总动量 total momentum
总透射比 total transmittance
总线 bus
总星系 metagalaxy
纵波 longitudinal wave
纵模 longitudinal mode
纵向弛豫 longitudinal relaxation
纵向放大率 longitudinal magnification
纵向像差 longitudinal aberration
纵质量 longitudinal mass
纵坐标 ordinate
族 family; group
阻碍因子 hindrance factor
阻挡层 barrier
阻光性 opacity
阻抗 impedance
阻抗匹配 impedance matching
阻力 resistance
阻尼 damping
阻尼力 damping force
阻尼碰撞 damped collision
阻尼系数 damping coefficient
阻尼谐运动 damped harmonic motion
阻尼振荡器 damped oscillator
阻尼振动 damped vibration
阻尼振子 damped oscillator
阻容耦合放大 amplification by R–C coupling
阻塞效应 blocking effect
阻止本领 stopping power
阻止距离 stopping distance
组分 component
组合 combination
组合宇称 combined parity
组合原理 combination principle
组态 configuration
组装密度 packing density
最大磁能积 maximum magnetic energy product
最大偏差 maximum deviation
最大容许剂量 maximum permissible dose
最大似然法 maximum likelihood method
最大误差 maximum error
最大值 peak (value)
最低未占分子轨道 LUMO (lowest unoccupied molecular orbit)
最低温度计 minimum thermometer
最概然分布 most probable distribution
最概然速率 most probable speed
最概然值 most probable value
最高温度表 maximum thermometer
最可几速率 most probable speed
最可几值 most probable value
最小二乘法 least-squares method
左手定则 left-hand rule
左旋(光) levorotation
左旋晶体 left-handed crystal
左旋形式 l-form
左旋圆偏振 left-hand circular polarization
左右倒置 lateral inversion; perversion
作用 action
作用光谱 action spectrum
作用力 acting force
作用量 action
坐标 coordinate
坐标几何 coordinate geometry
坐标系 coordinate system; frame of axes

非汉字开头的词条

BCS 理论　BCS theory
b 夸克　bottom quark；b-quark
c. g. s. 单位制　c. g. s. system
CPT 定理　CPT theorem
CP 不变性　CP invariance
CP 破坏　CP violation
CP 宇称　CP parity
c 夸克　c-quark
C 宇称　C-parity
d 夸克　d-quark
E 层　Heaviside layer
F 电离层　Appleton layer
f 数　f-number；focal ratio
GWS 模型　GWS model（Glashow-Weinberg-Salam model）
g 态　g-state
G 宇称　G-parity
J/ψ 粒子　J/psi particle
j-j 耦合　j-j coupling
KAM 定理　KAM theorem（Kolmogorov-Arnold-Moser theorem）
KAM 环面　KAM torus
K 介子　kaon
L-S 耦合　L-S coupling；Russell-Saunders coupling
MOS 场效管　MOSFET（metal-oxide-semiconductor field effect transistor）
MOS 晶体管　MOS transistor（metal-oxide-semiconductor transistor）
n 区　n-region
n 型半导体　n-type semiconductor
n 型电导率　n-type conductivity
pH 值　pH
p-n 结　p-n junction
p-n 结二极管　p-n junction diode
p-n 结激光器　p-n junction laser
p 型半导体　p-type semiconductor
Q 开关　Q-switch
q 数　q-number
Q 调制　Q-modulation
Q 突变　Q-spoiling
Q 因数　Q factor（quality factor）
Q 值　Q-value
S 波　secondary wave
S 矩阵　S-matrix（scattering matrix）
S 矩阵理论　S-matrix theory
S 型熔丝　type-S fuse
T 不变性　T-invariance
T 破坏　T violation
U 形管　U-tube
WS 模型　WS model（Weinberg-Salam model）
W 玻色子　W-boson
W 粒子　W particle
X 光　X-ray
X 射线　X-ray
X 射线管　X-ray tube
X 射线激光　xaser；X-ray laser
X 射线激光器　xaser；X-ray laser
X 射线激射器　xaser；X-ray laser
X 射线晶体学　X-ray crystallography
X 射线谱　X-ray spectrum
X 射线天文学　X-ray astronomy
X 射线衍射　X-ray diffraction

X 射线荧光 X-ray fluorescence
YAG 激光器 YAG laser
Z 玻色子 Z boson
Z 粒子 Z particle

1/4 波片 quarter-wave plate

α 放射性 alpha activity；alpha radioactivity；α-radioactivity
α 辐射 alpha radiation
α 粒子 alpha particle
α 粒子-氘核模型 alpha-deuteron model
α 粒子计数管 alpha counter tube
α 射线 alpha ray；α-ray
α 衰变 alpha decay；α-decay
α 铁 alpha-iron
β 放射性 beta radioactivity；β-radioactivity
β 辐射 beta radiation
β 粒子 beta particle
β 射线 beta ray；β-ray
β 衰变 beta decay；β-decay
β 铁 beta-iron
γ 放射性 gamma radioactivity；γ-radioactivity
γ 辐射 gamma radiation
γ 射线 gamma ray；γ-ray
γ 射线天文学 gamma-ray astronomy
γ 衰变 gamma decay；γ-decay
θ 真空 theta vacuum；θ vacuum
λ 点 lambda point
λ 粒子 lambda particle
μ 子 mu-meson；muon
π 电子 pi-electron
π 介子 pi-meson；pion
σ 粒子 sigma particle
τ 粒子 tau particle
τ 轻子 tau lepton
ψ 粒子 psi particle
Ω 负粒子 omega-minus particle

参考文献

程希望等编. 英汉光学与光通信词汇. 北京：科学出版社，2004.

《力学词典》编辑部编. 力学词典. 北京：中国大百科全书出版社，1990.

倪光炯等. 改变世界的物理学. 上海：复旦大学出版社，1998.

汪德昭主编. 英汉声学词汇. 北京：科学出版社，1982.

《新英汉词典》编写组编. 新英汉词典(增补本). 上海：上海译文出版社，1985.

《英汉综合物理学词汇》编辑组编. 英汉综合物理学词汇. 北京：科学出版社，1999.

张缵绪主编. 英汉数理化词典. 北京：中国标准出版社，1991.

赵凯华主编. 英汉物理学词汇. 北京：北京大学出版社，2002.

中国社会科学院语言研究所词典编辑室编. 现代汉语词典. 第5版. 北京：商务印书馆，2006.

祖贝编. 英德法俄汉物理学词典.《物理学词典》翻译组译. 北京：原子能出版社，1979.

Isaacs A. *Oxford Dictionary of Physics*. 3rd ed. Oxford: OUP, 1996.

外教社英汉·汉英百科词汇手册系列

1. 保险
2. 材料学
3. 财政学
4. 测绘学
5. 出版印刷
6. 地理学
7. 地球科学
8. 地质学
9. 电子、通信与自动控制技术
10. 电子商务
11. 动力与电气工程
12. 法学
13. 房地产
14. 纺织
15. 工程与技术
16. 管理学
17. 广播影视
18. 广告学
19. 国际贸易
20. 海洋与水文
21. 航空航天
22. 核科学技术
23. 化工
24. 化学
25. 环境科学
26. 机械工程
27. 计算机
28. 交通运输工程
29. 教育学
30. 经济学
31. 军事
32. 考古学
33. 会计与审计
34. 矿山工程技术
35. 力学
36. 历史学
37. 林业
38. 旅游
39. 美术、书法与摄影
40. 民族学
41. 能源科学技术
42. 农业
43. 烹饪
44. 人口学
45. 人类学
46. 人力资源管理
47. 商务
48. 社会学
49. 生理学
50. 生物学
51. 食品科学
52. 市场营销
53. 市政工程
54. 数学
55. 水产业
56. 水利工程
57. 体育
58. 天文学
59. 统计学
60. 图书馆、情报与文献学
61. 土木建筑
62. 网络与多媒体
63. 微生物与病毒学
64. 文学
65. 物理学
66. 物流与仓储
67. 戏剧、戏曲与舞蹈
68. 心理学
69. 新闻与传播
70. 信息与系统科学
71. 畜牧与兽医
72. 药学
73. 医学
74. 音乐
75. 语言学
76. 园艺学
77. 哲学
78. 政治学
79. 知识产权
80. 植物学
81. 中医中药
82. 宗教